Pearson's Canal Companion
LEEDS & LIVERPOOL
West Yorkshire Waterways

Published by Wayzgoose
Staffordshire DE13 9RS
email: enquiries@jmpearson.co.uk
www.jmpearson.co.uk

WAYZGOOSE

Dog-walking, Ferrybridge (Map 9)

LOCK-WHEELING

Regrettably, I wasn't able to attend my seventieth birthday party. Characteristically, I was working. Very time-consuming, the Canal Companions. *Very*! But, taking my cue from a Monty Python sketch, lost in the impenetrable mists of comedic time, I sent my fridge to represent me; rammed with Saltaire Blonde and Leventhorpe Brut, so that the assembled throng could hardly quibble. Besides, they were much better off without a party-pooper present. For, to purloin a line from an old Paul Simon song: 'I'm not the kind of man who tends to socialise.'

Just as one is never fully grown at thirty, neither is one fully old at seventy. The aspirations and ambitions of youth remain fundamentally intact; though opportunities to accomplish them are diminishing, hour by slippery hour. One embarks on journeys one is not guaranteed to complete. Undertaking this update, I was forcibly reminded of this sobering verity at East Marton, where the Leeds & Liverpool Canal briefly rubs shoulders with the Pennine Way. Back in the early Seventies, Alfred Wainwright was a hero of mine, and my well-thumbed, pork pie-crumbed second impression of his *Pennine Way Companion* accompanied me on a dry run from Alston to Hawes, subliminally instilling virtuous guide book compiling techniques in me as I went. Thus it was gratifying - flattering, even - decades later, to be likened by one Amazon reviewer to: 'the Wainwright of the Waterways'. If only I'd been given my own television show! If only I'd amassed his royalties! If only I'd got to go on Desert Island Discs! Well, still time; though as likely as a Liberal government.

*   *   *

'Ey-oop, stranger!' says The North, in that bluff vernacular popularly, if unfairly, ascribed to it. 'What brings tha' back to these parts?' And I, in that ingratiating tone the midlander - clearly at an interlocutory disadvantage - feels compelled to adopt when faced with the cardinal points of the compass, respond to the effect that it's well overdue a post-pandemic update, the equivalent, perhaps, of the sort of medical check-up the NHS goes in for these days, whether one has any symptoms or not.

'And how's tha' found us?', The North enquires, archly.

'Oh, as adorable and as inscrutable as ever,' I laugh.

Travelling back up the M1 after the Lower Sixth had been frog-marched to a performance of *Macbeth* at the Nottingham Playhouse, 'Boris' Atkinson and I amused ourselves - if palpably no one else present - by elaborating on signs for 'The North', conjecturing if the next would read: 'Further North'; and the one after that: 'Even Further North', or: 'You're Nearly North Now'. Then, going off at a tangent, so to speak, we idly wondered if there'd be a junction for 'Magnetic North'. Oh how we laughed!

Alighting on the stroke of midnight, moments before Broadbent's 'luxury' coach transmogrified into a pumpkin, 'Boris' and I surmised, somewhat precociously, that The North wasn't so much a destination, as a state of mind. I trust this guide to some of its watery parts proves as inspirational as my Wainwright was to that callow, twenty year old me.

Contents

Boatyard & Bottle Works, Knottingley (Map 9)

CALDER & HEBBLE Sowerby Bridge 3½mls/5lks/2hrs*

STUFF Santiago de Compostela. If you make one pilgrimage in your life, make it from Sowerby Bridge to Selby. Across the lost Romano-British Kingdom of Elmet. Preferably (though not mandatorily) by boat. From the Pennine foothills to the Plain of York you will encounter a scintillating slice of northern Eng and likely to stay with you long after more obviously touristy climes have vanished in a mental haze of Albarino.

There is one drawback, however, in commencing a pilgrimage from Sowerby Bridge, for it is such a characterful and atmospheric location - especially the canal basin - that you are inclined to come up with a string of excuses for delaying departure. One of the inland waterway system's great transhipment centres - another Shardlow, another Stourport; though more in the 'northern' mould - it was here in the early 19th century heyday of Trans Pennine carriage by water that Yorkshire keels exchanged cargoes with Mersey flats: not simply because no true Tyke could ever entirely trust someone from 't'other side o' Pennines', but because, at 70ft in length,

the Lancashire boats were too lengthy to negotiate the Calder & Hebble's stumpy locks. Gauge conflicts such as this, did little to help the canals compete with the emerging railways. Trade was abandoned across the summit of the Rochdale Canal before the Second World War. The last waterborne cargo to reach Sowerby Bridge from the east was paper pulp aboard the keel *Frugality* in September 1955.

Much refurbished of late, the basin nevertheless retains a good deal of the latent atmosphere of trade. Several large warehouses, together with small buildings such as a weighbridge, porter's lodge and agent's house, remain intact, combining to picturesque effect. These are overlooked by a massive Methodist church which once had its organ bellows powered by a water turbine. Everywhere you look there are charming cameos to admire, not least the life-size figures of a boatman and a boy putting pressure to a lock beam which welcomes folk into the basin off Wharf Street. Not even a fortune-teller would have dared suggest that, *continued overleaf:*

*figures relate to section east of Sowerby Bridge

continued from page 5:

forty years after *Frugality* docked, the Calder & Hebble and Rochdale canals would be relinked, but that was the almost miraculous turn of events which occurred in 1996. Once again you can boat across the Pennines to Manchester as exuberantly described in *Pearson's Canal Companion to the Cheshire & South Pennine Rings*. But this guide heads East, old man, and it's time you pilgrims girded your collective loins for pastures new.

The first bridge out from the basin used to have a chain drawn across it on Sundays to prevent boatmen moving boats on the Sabbath. The street it carries continues down to cross the Calder on a cast iron bridge built at the Shelf Iron Works in 1816, making it one of the earliest of its type in the North. Clinging to a shelf above the Calder, and cuddled in shallow hawthorn cuttings, or raised above the river on sapling-framed embankments, the inland waterway traveller's perspective is curtailed, creating a sense of intimacy that intensifies one's relationship with the canal.

The hills and mills and railway lines provide reminders of the Upper Peak Forest Canal, but the most spectacular sight on this length is Wainhouse's Tower, a pencil-thin 253ft high tower of blackened ashlar stone, erected as a chimney for a dyeworks, though never used as such and later converted into a viewing tower. Open to the public on selected dates, 403 steps will take you to the top where, they always said, you could see Blackpool on a clear day, but days were seldom clear in the smoky West Riding of years gone by, so you'd have to make do with Halifax.

The demolished mill at Sterne Bridge belonged to the Sterne family of novelist Laurence fame, writer of that rambling 18th century novel *Tristram Shandy*. Another literary connection here relates to Wordsworth's poem *Lucy Gray*, which was inspired by an incident involving the drowning of a young girl when the bridge was swept away by a winter flood. At the end of the poem, Wordsworth suggests that Lucy Gray is still to be seen: 'smoothly tripping along, singing a solitary song, and whistling in the wind'. West Yorkshire Constabulary's advice is that she shouldn't be approached.

Between Sterne Bridge and Salterhebble, the canal rides above a valley floor littered with sewage plants and factory premises. The Calder's flanks, however, are sylvan enough to create a favourable impression with those who like their pilgrimages on the pretty side. A many arched stone viaduct carries the Manchester-Bradford railway across the Calder & Hebble at Copley, a community based on a model mill village of 19th century worker's housing down by the riverside. The village - which predated Saltaire (Map 17) - was erected by an altruistic worsted mill owner called Edward Akroyd. The community's impressive French Gothic church, St Stephen's, had allegedly to be sited on the far side of the river to appease the Rector of Halifax. No longer used for regular worship, it is cared for now by the admirable Churches Conservation Trust, and contains 'a noteworthy sequence of stained glass by Hardman in the apse': Keyholder nearby! Bosky footpaths thread the neighbouring woods. Old Rishworthians play rugger at Copley and there is a splendid little cricket ground boxed in by the railways and the river - do they award *seven* runs if you manage to hit the ball over the viaduct?

The trio of locks at Salterhebble are unarguably the most photogenic and well cared for on the Calder & Hebble. They are laid out around a dog-leg curve. The top two were once a staircase, whilst the bottom lock features an unusual, electrically operated guillotine gate at its tail, an introduction necessitated by the widening of the adjacent road bridge. We were interested to see that the gate was manufactured by Ransome & Rapier of Ipswich, a firm perhaps better known, in transport enthusiast circles at least, for their railway turntables.

In the short pound between the middle and lower locks, a small aqueduct spans the Calder & Hebble's junior partner; a shy, retiring sort of chap called Hebble Brook, who springs to life on t'moors above Halifax. Time was when a branch canal climbed through fourteen locks in less than two miles to reach the centre of that sooty, hill-bounded textile and engineering town. Even as canals go, it was a heavy drinker. And because the local mills already had the valley's water supplies sewn up, water for the branch was pumped (at the rate of a thousand gallons a minute) up from the main line in an Escher-like arrangement of perpetual motion. Opened in 1828, the Halifax Branch became one of the less publicised abandonments of the LMS Railway's infamous 1944 Act, and much of its course has subsequently been obliterated, though you get some idea of its topography from the right hand side of a slow-climbing train, or more intimately by walking or cycling the Hebble Trail, an entertaining, traffic-free right of way between the foreshortened terminus of the branch and the centre of Halifax. It amused us to discover that one of the branch's more notable cargoes was goux, a form of sewage. We had always considered this a colloquial euphemism until we discovered that it derives eponymously from Pierre Nicolas Goux, a 19th Century Parisian who developed a system of domestic sewerage involving pails lined with absorbent material which, as far as England was concerned, caught on almost exclusively in Halifax. Goux was taken along the Calder & Hebble to the East Riding and Lincolnshire for use as fertilizer. A short stub of the Halifax Branch remains in water at the Salterhebble end, with a Brewers Fayre, Premier Inn, McDonald's, and frequent buses into Halifax (you'd be foolish to miss the iconic Piece Hall while in the neighbourhood) providing reason enough for a short detour with room to turn at the end. The towpath side is given over to CRT long term moorings, but visitor moorings are available by arrangement with the receptionist at the adjacent Premier Inn for a not unreasonable fee donated to charity.

Couched between the busy A629 and River Calder, but well screened by woodland, the navigation makes picturesque progress, encountering two locks and passing beneath the old Lancashire & Yorkshire Railway.

Sowerby Bridge — Map 1

Squeezed into the precipitous Calder Valley - like a cork about to pop out of a bottle - 'Sorby' Bridge fizzes with atmosphere, and there are plenty of champagne moments for folk of a Pearsonesque sensibility who warm to the more saturnine corners of this sceptred isle. It's a great strapping Yorkshire lass of a place, prone to glower on first acquaintance, but increasingly wreathed in self-deprecating smiles as intimacy grows. For a taster try the covered alley which slips unnoticed away from Wharf Street by the fish & chip shop. A footbridge (offering illuminating views of the town's posterior) carries you across the Calder to the railway station (scene of an unnerving denouement in the second series of Sally Wainwright's BBC drama *Happy Valley*) before making your way down Station Road to the A58 and the river bridge, returning full circle via Town Hall Street. See if you can't time your arrival to coincide with the annual Rushbearing Festival - a boisterous bacchanalia of traditional processions, morris dancing, brass bands, jazz and folk music - which consumes the town on the first weekend in September.

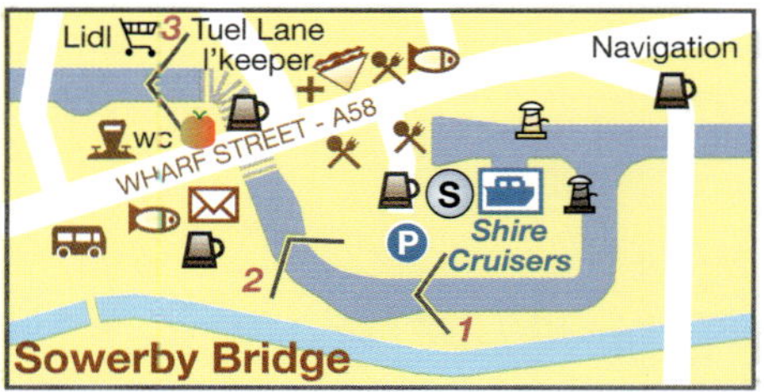

Eating & Drinking

ENGINE - Wharf Street. Tel: 01422 740123. Contemporary, 'small plates' restaurant, open Wed-Sat 12-9pm. HX6 2AF

GIMBALS - Wharf Street. Tel: 01422 839329. Once exceptional restaurant currently only operating for pre-ordered (by Thur 7pm) local deliveries Fri & Sat 2-5pm, or as a mouth-watering food shop on Friday and Saturday between 10am and 4pm. HX6 2AF

OLIVE BRANCH - Town Hall Street. Tel: 01422 832271. Turkish restaurant which serendipitously also has a branch in Selby, adding an extra dimension to our recommended pilgrimage - see page 5! HX6 2EA

NAVIGATION - Chapel Lane. Tel: 01422 750947. Post flood refurbished canalside pub. Food served 3-8pm Mon-Thur, and from noon Fri-Sun until 7pm (4pm Sun). Choice of Yorkshire ales. HX6 3LF

SYHIBA - Wharf Street. Tel: 01422 835959. Excellent town centre Indian restaurant. HX6 2AF

TEMUJIN - canal basin. Tel: 01422 835500. Mongolian cooking. Sink your teeth into crocodile, ostrich or kangaroo. HX6 2AG

VILLAGE - Wharf Street (opposite canal basin). Tel: 01422 831654. Asian cuisine. HX6 2AF

Shopping

Lots of useful outlets. Lidl handily placed just up the road from Tuel Lane Lock. Large Tesco across the river. Market on Tue, Thur, Fri & Sat by Tuel Lane Lock.

Connections

TRAINS - Calder Valley services from station easily reached by footpath and alleyway adjacent to fish & chip shop on Wharf Street. Tel: 0345 748 4950.

BUSES - services throughout Calderdale. Service 577 provides a hair-raising, circular, switchback ride up on t'moors. Tel: 0871 200 2233.

TAXIS - A1. Tel: 01422 835050.

RUBBING off on you yet, the Calder & Hebble's elusive charm? They don't make inland waterways like this down south! Venerable stone mileposts count down to faintly amusingly named Fall Ing at Wakefield where the C&H hands over to its uncle, the Aire & Calder: up north they like to keep things in the family. Atmosphere oozes from every pore of this unjustly undersung waterway: partly because of its unusual gauge (the stumpy locks measure 57ft 6ins x 14ft 2ins, and optically appear very squat indeed); partly because its course lies couched in such a fascinating landscape - half rural, half post-industrial - where interest seldom falters. Too complex to even hint at here, the navigation's history can be traced back to the middle of the 18th century and the West Riding woollen trade's aspirations of building on the success of the earlier Aire & Calder Navigation. Several A-list engineers of that halcyon canal-building era were at one time or another associated with its construction: Smeaton, Brindley, Jessop et al.

River rarely out of sight, the navigation proceeds through the Calder Valley accompanied (west of Brighouse) by a metalled towpath designated the Calder Valley Greenway. Locks appear as frequently as unsolicited litigation calls, reflecting a plethora of weirs on the river itself - in its twenty-two miles the C&H descends the best part of two hundred feet between Sowerby Bridge and Wakefield.

The characterful little town of Elland looms over the navigation as the towpath swaps sides, involving a short but charming detour to the rear of the handsome wharf buildings, headquarters of William Dutton, one of the Calder & Hebble's principal carriers before being bought out by Hargreaves upon Nationalisation in 1948. Elland was the home of the Gannex mac, famously modelled by the pipe-smoking Prime Minster, Harold Wilson. Elland Lock retains its keeper's office, as does Brookfoot further down the valley. A two-chimnied salute rises above the treeline from

In the absence of a riverside path, walkers will need to detour from the route the river takes between Brighouse and Anchor Pit Flood Lock. At least they gain some insight into what makes Brighouse's economy tick.

the premises of W. T. Knowles, one of the last manufacturers of clay pipes and chimney pots.

At Brookfoot Lock there is evidence of an old link with the river, used until the middle of the 20th century to access riverside sand and gravel pits, indeed, sometimes vessels would dredge sand from the bed of the river itself. The towpath changes sides again at Ganny Lock, squeezed between a rocky cliff face and the broadening Calder. At Brighouse, the Calder - which has been beseeching the canal to let it play with it - is finally allowed to join in the fun and games ... and when the river is at its most mercurial, what games they can be. Do not enter the river if the level gauge shows only red. The town's waterfront is dominated by a pair of disused silos, which lend it character, if not a mildly enervating post-industrial despair; and yet, Sugden's Brighouse Mill has found an imaginative new lease of life as a 'climbing gym'. Sugden was a builder of keels before becoming a dusty miller. Wheat was still being brought by barge to the mill when The Beatles first appeared in the hit parade. Anchor Bridge retains its old triangular number plate 'C&H 9'. Not so the following bridge, Huddersfield Road, which British Waterways (before being reinvented as CRT) mysteriously renumbered 18. Brighouse Basin is overlooked by Mill Royd Mill which has been transformed into apartments.

The navigation drops down into the river at the tail of the lower of the two Brighouse Locks. Before it was swept away by floods in 1946, there was a hauling bridge here, enabling boat horses to reach the towing path on the south bank of the river. The Calder injects an element of whimsy - at least where boaters are concerned - down to Anchor Pit. Flood locks are a feature of the Calder & Hebble and Aire & Calder navigations. Some are merely single pairs of mitred gates employed to keep excess water out of the canalised cuts at times of high water levels. Others have top and bottom gates and can be made to operate as a conventional pound lock. On the bend before Anchor Pit Flood Lock, two rows of terrace housing still sport outside privies, though one imagines they are more likely to be used as ancillary domestic repositorys now than for calls of nature.

Elland
Map 2

Characterful small town perched above the Calder. A business park has replaced the mills. The bridge dates back to the 17th century, though with subsequent lengthenings and widenings. Facing it, an imposing facade, supported by four Aberdeen granite columns and topped by a seated Neptune, was originally the august premises of the Halifax & Huddersfield Bank. Notable businesses include Dobson's boiled sweet factory. Make a point of attending the impossibly quaint Rex Cinema where the Rodgers 333 Olympic theatre organ is still played at recitals on the third Sunday in the month.

Eating & Drinking
BARGE & BARREL - Park Road (canalside Elland Bridge). Tel: 01422 254604. HX5 9HP
CRAFT & TAP - Southgate. Tel: 01422 377677. Elland Brewery micropub. Open Mon-Thur from 3pm; Fri & Sat from noon; Sun 2-8pm. Bar snacks. HX5 0EP

Shopping
Elland provides good shopping facilities (including a large Morrisons) and gravity should aid your package-laden progress back to the boat. Dobson's charming sweety shop lies on Southgate.

Connections
BUSES - in the regrettable absence of the trains, trams, or trolley-buses of yore, bus services link Elland with the rest of Calderdale. Tel: 0871 200 2233.
TAXIS - Elland Cars. Tel: 01422 327444.

Brighouse
Map 2

Best known for its brass band, Brighouse is an engaging little town with a network of largely traffic-free streets made up of shot-blasted clean Victorian stone buildings. Solely in Pearson's would you glean that Wagner's grand-daughters went to school here.

Eating & Drinking
BELLINI'S - Phoenix Street. Tel: 01484 400114. Italian restaurant from 5pm ex Mon. Easily reached just north of basin. HD6 1PD
BLAKELEY'S - Canal Street. Tel: 01484 713907. Good fish & chip restaurant and take-away. HD6 1JX
BROOK'S - Bradford Road. Tel: 01484 715284. Highly regarded restaurant. Closed Mon & Tue. HD6 1RW
MARKET TAVERN - Ship Street. Tel: 01484 769852. Cosy/friendly micropub. Closed Mon/Tue. HX6 1JX
RED ROOSTER - Elland Road. Tel: 01484 968402. Re-opened real ale local serving pies & peas. Open from 2pm Mon-Thur and from noon Fri to Sun. HD6 2QR

Shopping
The best reason for shopping in Brighouse continues to be Czerwick's wine & cheese (and Andrew Jones pies) shop on Commercial Street, followed closely by Le Gourmet pork pie shop & delicatessen on Bethell Street. Market on Wednesdays and Saturdays.

Things to Do
SMITH ART GALLERY - Halifax Road. Tel: 01422 288065. Enjoyable little gallery bequeathed by local Victorian manufacturer of tweeds and serge. Collection includes an Atkinson Grimshaw. HD6 2AD

Connections
TRAINS - approx hourly links to Huddersfield, Halifax and Leeds (via Dewsbury). Tel: 0345 748 4950.
BUSES - services E6/7 run to/from Elland; X6/63 to/from Huddersfield and Halifax. Tel: 0871 200 2233.
TAXIS - Woods. Tel: 01484 400800.

LUSTILY humming *Who Would True Valour See*, with Bunyanesque conviction, inland waterway pilgrims continue their way through the Calder Valley in high spirits, the friendly little town of Mirfield marking the halfway point between Sowerby Bridge and Wakefield or vice versa.

Kirklees Cut seems remote from the outside world. Minor artefacts evoke the past: distance posts indicating 100 yards to the locks; Lancashire & Yorkshire Railway boundary posts; and a curious horizontal pulley wheel which would have been used to give boat horses extra purchase when starting away from the lock. Robin Hood and Little John forded the Calder in 1247 at the end of their ill-fated journey to Kirklees Priory, where the hero of Sherwood Forest is said to have been bled to death by the treacherous abbess.

At Cooper Bridge the waters of one of the Calder's most significant tributaries, the Colne, make their presence felt amidst a plethora of sewage works. Good visitor moorings are obtainable in the lock cut where you are allowed 72 hours to acclimatize yourselves to the fruity aroma of the neighbouring filter beds. Collectors of the obscure and enigmatic should stroll down the A62 to its junction with the A644 to see a charming little curiosity called the Dumb Steeple.

Either side of Cooper Bridge, the Manchester & Leeds Railway of 1840 crosses the navigation twice in quick succession, its original stone bridges being supplemented by iron girders when the line was quadrupled circa 1900. Mirfield Monastery - occupied by the Community of the Resurrection - stands aloof from the valley floor, its pale green rooftops a prominent landmark. The church was built between 1910 and 1939 on the site of a former quarry. Members of the community take vows of Stability, Obedience and Conversion of Life, concepts explained in more detail on the community's website (*www.mirfield.org.uk*). Bed & breakfast and retreats are available. Dietrich Bonhoeffer - who was executed by the Nazis in the wake of the failed attempt to assassinate Adolf Hitler - visited Mirfield in the 1930s. The house beside the monastery belonged to a family whose fortune was made selling blankets during the Franco-Prussian War. Bedding is still manufactured locally at the Nunbrook Mills which overlook the unnavigable stretch of the river.

Above Battyeford Lock the South Pennine Boat Club occupy the site

The riverside path between Kirklees Lower Lock and Cooper Bridge is increasingly eroded. Follow the footprints and you shouldn't go far wrong.

1: Mirfield Marina
2: Shepley Bridge Marina

1: Cooper Bridge - csd 1950
2: Battyeford - csd 1953

NB: no access to Old Mill pub from Kirklees Locks

of a former boatbuilding yard whose drydock survives, enthusiastically employed by members to blacken their bottoms.

A river section ensues, formerly spanned by an impressive Warren Truss girder bridge which carried the London & North Western Railway's 'Leeds New Line'. Opened in 1900, the line only lasted for sixty-five years and the bridge was removed for scrap in 1967. A second bridge has vanished from this reach of the Calder, though it has absentmindedly left its central pier behind. A potential hazard for negligent boaters, and a sorely missed facility for anglers and towpath walkers, the bridge was originally provided to enable boat horses to cross the river. It was demolished in 1958.

A large weir precedes Ledgard Flood Lock where downstream boaters pass onto the calmer waters of the Mirfield Cut. Memories of Mirfield as an industrial centre are hazy now, but the cut was once lined with maltings and a notable boatbuilding yard where the 'West Country' keel *Isobel* - reputedly the last wooden barge to be built for commercial service on a British waterway - was launched sideways with a big splash in 1955. An interpretive board alongside the visitor moorings illustrates the event, and also outlines the history of Crowthers, the last survivor (albeit then owned by Bass Charrington) of several maltsters based in the town. Textiles were

another significant component of the local economy, as were the railways; there was a motive power depot here (on a site now covered by housing), whilst the station waiting room apparently once boasted the luxury of a billiard table for passengers to while away the hours between trains. Happily, the Mirfield Boat Company maintain the tradition of boatbuilding from premises between Bull and Gill bridges, constructing both broad and narrowbeam steel craft for the leisure market.

The stone abutments of a dismantled bridge which once carried the Lancashire & Yorkshire Railway's Cleckheaton branch across the cut precede Wheatley Bridge and Shepley Bridge Lock where there was another boatbuilding yard, nowadays a small marina. Below the lock the Calder returns, as if itching to get in on the action, but its impact is brief and it goes muttering away to itself over a weir, as you enter the motherly embrace of Greenwood Cut. Ravensthorpe was the western terminus of the delightfully titled Yorkshire Woollen District Electric Tramways. A proposed extension of the Huddersfield Corporation Tramways to meet at this point never materialised, though a cast iron post on the roadside reads 'Terminus'. Dewsbury Bus Museum lies about 10 minutes walk along the A644 - see Map 4.

Mirfield Map 3

Map 3

The mills have become apartments and the railway yards housing, whilst the sites of maltings have been colonised by supermarkets. Yet Mirfield - back in the Heavy Woollen District smogs of time a market town of some stature - retains a fragile sense of individuality and tenuously boasts connections with the Brontes. All three sisters were educated at Roe Head school on the moors to the north-west of the town in the 1830s; Charlotte going on to teach there. In turn, Anne became a governess at Blake Hall (demolished 1954), an experience which is said to have inspired her novel *Agnes Grey*. St Mary's imposing church is largely the work of George Gilbert Scott.

Eating & Drinking

ALEXANDER'S - Huddersfield Road. Tel: 01924 497039. Bar & kitchen open from 4pm Mon-Fri and from noon at weekends. WF14 8AB
FLOWERPOT - Calder Road. Tel: 01924 496939. Ossett Brewery pub south of the railway station alongside the unnavigable Calder. WF14 8NN
NAVIGATION TAVERN - canalside Station Road. Tel: 01924 492476. Homely inn, with County Donegal connections, set back from canal bank. WF14 8NL
PEAR TREE INN - Huddersfield Road. Tel: 01924 491360. Cosy pub on the Calder at Battyeford. Bar/restaurant food. Moorings for patrons. WF14 9DL
SHIP INN - Shepley Bridge. Status uncertain.

Shopping

Lidl (alongside Bull Bridge), Co-op (incorporating Post Office), and Tesco Express. The West Yorkshire Print Workshop - a collective of local artists - is housed in an old chapel on Huddersfield Road. Gifts, books and crafts available at Mirfield Bookshop at the Community of the Resurrection on Stocksbank Road.

Connections

BUSES - services throughout Kirklees. Services 202/203 provide a half-hourly link to Leeds (via Dewsbury) and Huddersfield. Tel: 0871 200 2233.
TRAINS - frequent connections with Dewsbury, Brighouse & Huddersfield. Tel: 0345 748 4950.
TAXIS - Central Cars. Tel: 01924 490000.

DON'T you find it odd that the Calder & Hebble seems to cold shoulder Dewsbury, the largest town it passes through? It wasn't always thus. Wharves were established on the river as early as the 1760s as the C&H forged its way westwards. But the millers weren't enamoured, claiming that the navigation was extracting *their* water (as opposed to Nature's), and petitioned for a by-pass. Arbitration intervened, and the Long Cut was opened in 1798. Initially the original main line via Savile Town was abandoned. But it was re-opened - with a fine piece of one-upmanship - in 1865 by the Aire & Calder Co. who erected some splendid ware-houses on

the site, regrettably demolished in the 1970s as they were deemed a fire risk. The Old Cut was retained as well, along with a parallel stub called Fearnley's to serve industrial premises already established in the vicinity. You could even rent a rowing boat from underneath the gothic arches of the railway bridge and take your best girl for a dalliance past the symbolically high chimnied woollen mills of Ravensthorpe.

But all this historical scene-setting doesn't give you much of a flavour of the present day Calder & Hebble's traverse of Dewsbury's outlying suburbs. In truth, there is barely a dull moment. You can hear (subject to engine revs) West Yorkshire breathing beside you: hammering, crunching, belching, snorting; together with Tannoyed beseechments to report to reception. But down on the water you're mostly impervious to its post-industrial mood swings. Yours is the luxury of passing through. The Perseverance pub by Forge Lane Bridge seems aptly named. Aptly named, that is, until you nip up the steps in search of a quick morale-raising pint and find it dolefully boarded-up beyond redemption. Ditto

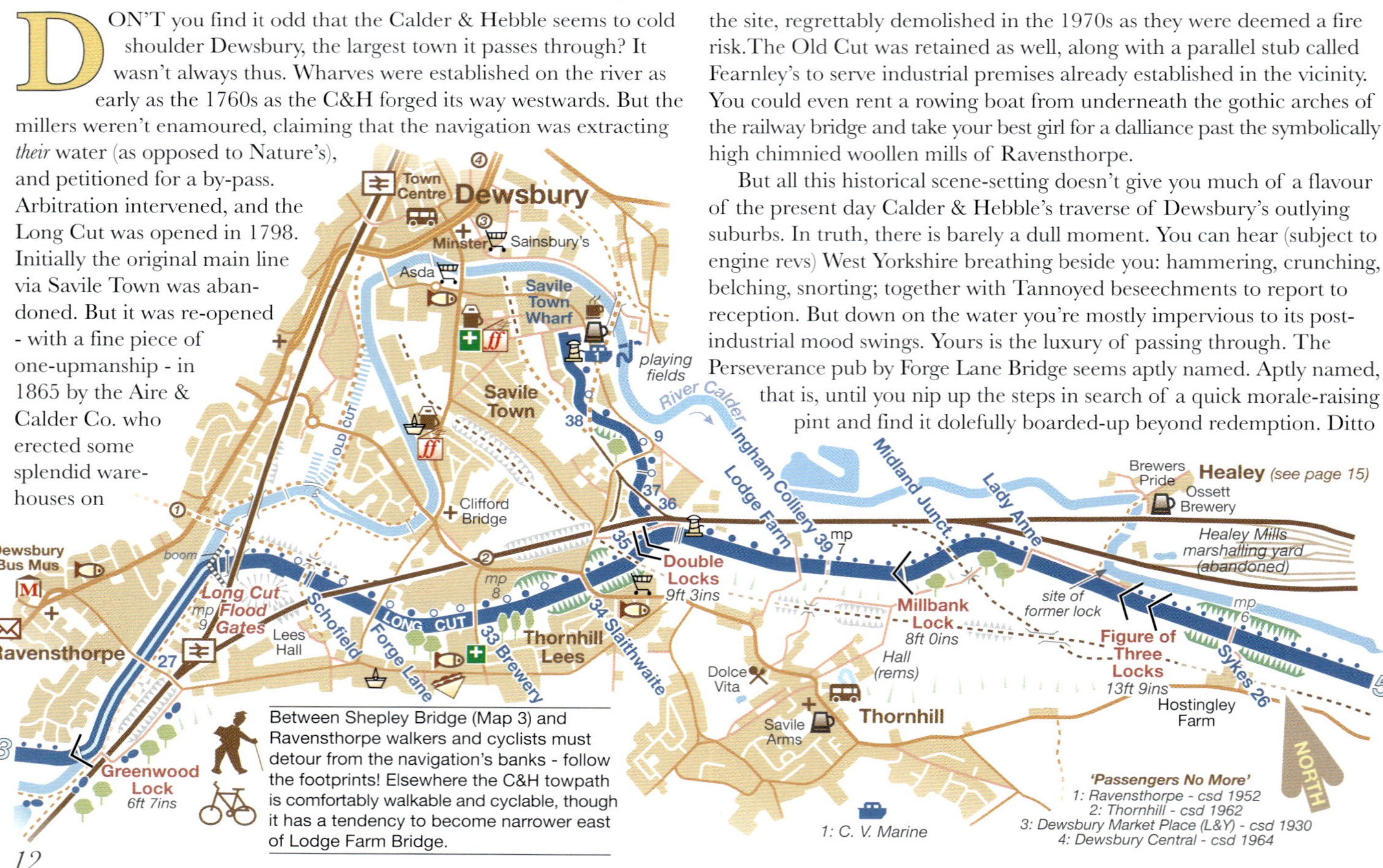

Between Shepley Bridge (Map 3) and Ravensthorpe walkers and cyclists must detour from the navigation's banks - follow the footprints! Elsewhere the C&H towpath is comfortably walkable and cyclable, though it has a tendency to become narrower east of Lodge Farm Bridge.

the Nelson by Bridge 34. Other pubs are available!

Meanwhile, back at Ravensthorpe, what is now known as Bridge 27 hangs skeletally over a broad stretch of the river lined on the north bank by residual pockets of mills interspersed with bland factory units, and on the south by Thornhill's now gas-powered electricty generating plant. Opened circa 1900, and much extended in the Thirties, it was the last user of water transport on the Calder & Hebble, Hargreaves *Haddesley* discharging the final load of coal from British Oak Staithe (Map 5) on 31st July 1981.

A handsome cast iron railway bridge dating from 1847 creates a portal to the Long Cut. Nature has untidily reclaimed Thornhill Iron & Steel Works, and new housing occupies the site of a brewery (though note the offside moorings rings), but lofty poplars (lending the C&H the look of a canal in Northern France) mask where the huge Providence Glass Works used to stand. In its heyday the works covered eleven acres and was churning out three hundred thousand bottles a week. Overlooking the cutting beyond Slaithwaite Bridge stood Hebble Mills, manufacturers of mungo and shoddy. High arches frame the approach to Double Locks, they belong to a railway viaduct (No.35) erected by the Midland Railway during their ill-fated Edwardian expansion in the West Riding.

Boaters on these northern backwaters are all explorers at heart, and will probably not be able to resist adding the Dewsbury Arm to their portfolio, any more than the Aire & Calder could. Why, indeed, should they? A rail served cement works provides a misleading character reference. Soon silver birch and water lilies create a more soothing aspect. A mural on a gable end harks back to trading days. Progressively narrower and reedy, the arm terminates alongside a Muslim High School, attended by exuberant girls filling the playground with Larkinesque 'whoops and skirls'. Visitor moorings are by arrangement with the neighbouring Calder Valley Marine who also provide long term moorings and boating facilities. Current bridge numbers bear no relation to the surviving triangular C&H plates, still demurely attached to some of the bridges in fading, air-gun-pellet-scarred, blue and yellow mortification. Neither are they necessarily in numerical order. Baffling!

East of the Double Locks, the Calder & Hebble enjoys a rural interlude, or what passes for rural in this fragmented West Yorkshire landscape. Figure of Three Locks are thought to have gained their curious name from the original convoluted shape of the River Calder hereabouts, rather than the fact that there was once a third lock connecting the navigation with the river. In February 2020, Storm Ciara damaged the locks so badly that they were closed for a year and £2 million had to be spent on their repair. Farms, which once existed cheek by jowl with collieries, now have the countryside to themselves. Geese fatten at Hostingley Farm.

Dewsbury Map 4

Once one of the old West Riding's 'big six' County Boroughs, Dewsbury's soot-cleaned Victorian and Edwardian architecture repays exploration.

Eating & Drinking

LEGGERS - Savile Wharf, Mill Street East. Tel: 01924 488153. Loft conversion of boat horse stables. Food and range of beers. WF12 9BD

OLD STABLES - Savile Wharf. Tel: 0739 657 3803. Cafe open Wed-Sun plus Thur & Sat eves. WF12 9BD

MINSTER REFECTORY - Tel: 01924 457057. Pleasant cafe open 10 am to 2pm ex Sat. WF12 8DD

WEST RIDING REFRESHMENT ROOMS - Wellington Road. Tel: 01924 459193. *Good Beer Guide* listed real ale pub in railway station waiting room. WF13 1HF

Shopping

Coach parties from across the north bear avidly down on Yorkshire's largest open market on Wednesdays and Saturdays, whilst throughout the rest of the week (early closing Tuesday excepted) the elegant indoor market hall of 1904 features a wide range of stalls. On the other hand, there is a large Asda supermarket less than ten minutes walk from Savile Town Wharf, and an Asian supermarket (Mullaco) near Bridge 34.

Things to Do

DEWSBURY BUS MUSEUM - Foundry Street, Ravensthorpe. Check internet for events. WF13 3HW

DEWSBURY MINSTER - Tel: 01924 466076. Handsome church which can trace its origin back to the ministrations of St Paulinus in AD 627. Irish born Patrick Bronte - father of Charlotte, Emily and Anne - was a curate here. WF12 8DD

Connections

TRAINS - Northern services to/from Leeds, Mirfield, Brighouse and Huddersfield. Tel: 0345 748 4950.

TAXIS - Royal Cars. Tel: 01924 450000.

ECHOES of lost endeavours abound as the Calder & Hebble traverses a region for the most part rural now, a scorched-earth zone which has successfully reseeded itself. That it resonates with Pearsons will come as no surprise, for we like our landscapes wounded, scarred and just a little bit haunted.

Used for residential moorings, Horbury Basin once effected another link with the Calder, sections of which remained navigable to serve riverside wharves after the Broad Cut was opened in 1838. In the 18th century a man-made channel to the north of Horbury Bridge by-passed a weir enabling craft to trade upstream towards Dewsbury. Beeston Bridge is numbered 28 now, but still bears an old CH numberplate 31: Hartley Bank Bridge and Waller Bridge being 32 and 33 respectively in 'old money'. Hartley Bank Colliery closed in 1968 and, frankly, you wouldn't know it had ever existed where green fields now run innocently down to the canal.

The elegant, tapered spire of St Peter & St Leonard's church in Horbury draws the eye to the north. It is the work of John Carr who fittingly lies buried there. He will be encountered again at Ferrybridge - Map 9.

A disused railway girder bridge hovers lugubriously over the water like a huge mechanical heron frozen in flight. Someone has expended a little tender shrine-like care on the site of British Oak colliery staithe from which the last cargoes of coal for Thornhill Power Station (Map 4) were despatched. Hargreaves had the contract for the run, and the ex British Railways 'Jinty' shunter which propelled the wagons down from the mine was painted in a gaudy shade of orange to match the livery of their barge fleet. How you long for loading to remain an everyday event. A lengthy blue brick viaduct carries the Midland Railway's ill-fated West Riding Extension across marshy ground to the south.

A short detour is required at Calder Grove where a cavernous footpath passes somewhat unnervingly beneath the Barnsley-Wakefield railway bridge.

Calder Grove is a popular resting place for local boaters; obviously they know just how good the fish & chips are up the road. Locally mined coal used to be loaded onto barges at Waller Bridge for despatch to York via the Aire & Calder, Selby Canal and River Ouse.

Hargreaves owned a parcel of land to the south of Broad Cut on which it was their habit to winch redundant wooden barges out of the water and set fire to them. A major employer in these parts was Charles Roberts, ostensibly a manufacturer of humble railway wagons, but not averse to turning their hand to tramcars and omnibuses. During the Second World War the plant was turned over to the production of tanks and munitions. Just about the last rail-based activity here concerned the fitting out of Virgin's Belgian-built 'Voyager' fleet. Nowadays the site has become a business park, but a blue plaque recalls the company's heyday.

A river section carries the navigation beneath the M1 Motorway as Emley Moor's transmitting mast - taller than the Eiffel Tower - becomes a dominant landmark to the south-west. This was an area of gravel extraction part of which has been reborn as Pugneys Country Park. Calder Park is quite a different thing altogether, an area devoted to business.

Thornes Cut eradicates one of the Calder's discursive meanders. Once upon a time Thornes Lock had duplicated chambers to cater for the sheer volume of barge traffic. An adjoining wire works featured a large asbestos clad warehouse with a canopy overhanging the river. Barges brought coils of wire here until 1967, but the works has been demolished and replaced - some might say symbolically - by a health & fitness club.

There are glimpses, to the south-east, of the motte & bailey remains of Sandal Castle, whilst nearby a housing estate occupies the site of the Battle of Wakefield in 1460, one of the rainbow-coloured reminders that "Richard of York gave battle in vain." The north bank of the river is sometimes given over to rhubarb growing (a local speciality) against a backdrop of Wakefield's skyline, dominated by the soaring spire of the city's cathedral, the art nouveau county hall and the high clock-towered French renaissance town hall.

Healey
Map 4

Little industrial community on the fringe of Ossett. Accessible from Figure of Three Locks by bridleway. Saplings thrive between the rails and floodlights no longer illuminate shunting through the small hours at the abandoned marshalling yard, opened in 1963.

Eating & Drinking
BREWERS PRIDE - Low Mill Road. Tel: 01924 273865. Ossett Brewery pub near the brewery itself. Open from 3pm Mon & Tue, otherwise from noon. A Blue Plaque records it as the childhood home of the youngest known service casualty in WW2. WF5 8ND

Things to Do
OSSETT BREWERY - Low Mill Road. Tel: 01924 261333. Brewers of growing influence throughout West Yorkshire. Shop open open 9am-4.30pm, Mon-Fri, tours by prior arrangement. WF5 8ND

Horbury Bridge
Map 5

The Reverend Sabine Baring-Gould was a curate here in 1864 and is said to have written the hymn *Onward Christian Soldiers* to encourage processions of children in the ascent of nearby Quarry Hill.

Eating & Drinking
CAPRI - adjacent Horbury Bridge. Tel: 01924 263090. Lively Italian bar/restaurant. WF4 5QA
Two other pubs, an Indian restaurant, and Chinese take-away across the river bridge.

Shopping
Post office stores to north and filling station with shop to south.

Things to Do
NATIONAL COAL MINING MUSEUM - Caphouse Colliery, Overton. Tel: 01924 848806. Open Wed-Sun 10am-5pm. Admission free. Station Coaches service 128 runs hourly Mon-Fri from Horbury Bridge past the entrance to the museum which is three miles to the south-west. Arriva 232 also provides an hourly service Mon-Sat. WF4 4RH

Calder Grove
Map 5

Eating & Drinking
THE NAVIGATION - canalside below Broad Cut Top Lock. Tel: 01924 274361. Stone-built inn divided into two distinct zones: a cosy lounge hung with archive photographs of the Calder & Hebble; and a restaurant/carvery. Large garden running down to waterside. Oh, and ice cream too! WF4 3DS
KINGFISHERS - Denby Dale Road. Tel: 01924 274994. Excellent, licensed fish & chip restaurant and takeaway, easily reached from the visitor moorings. WF4 3DA

Shopping
Newsagents and post office five minutes walk uphill.

PASSING from the Calder & Hebble to the Aire & Calder is like seeing a favourite old movie in widescreen for the first time. Or, to put it another way, the family resemblance is there, but the girth is altogether more substantial, as if the A&C exists solely on a diet of Yorkshire Curd Tarts and Taylor's Landlord bitter. Perhaps the bargemen of yore did, but they've been consigned to an increasingly not so recent history, and the waterway they worked on is tantamount to a funfair without dodgems now.

Still, we're here, so we might as well have some candyfloss, and you'd have to be of a singularly unimaginative disposition not to respond positively to Wakefield's re-emergent waterfront and the marvel that is Stanley Ferry Aqueduct.

Fields give way to factories as the Calder passes beneath the handsome railway bridge which carries the main line from London King's Cross to Leeds across the river. Its cast iron span, supported by battlemented abutments, is part of a lengthy viaduct known locally as the "ninety-nine arches", though some myth-shattering guidebooks would have it that there are only ninety-five. Did they *really* go and count them?

A notable riverside factory is that of W. E Rawson, carpet manufacturers, which was still receiving consignments of jute and coir fibre by barge in the mid Seventies. Nearby, modern double-deckers occupy garaging facilities once devoted to the upkeep of Wakefield's trams.

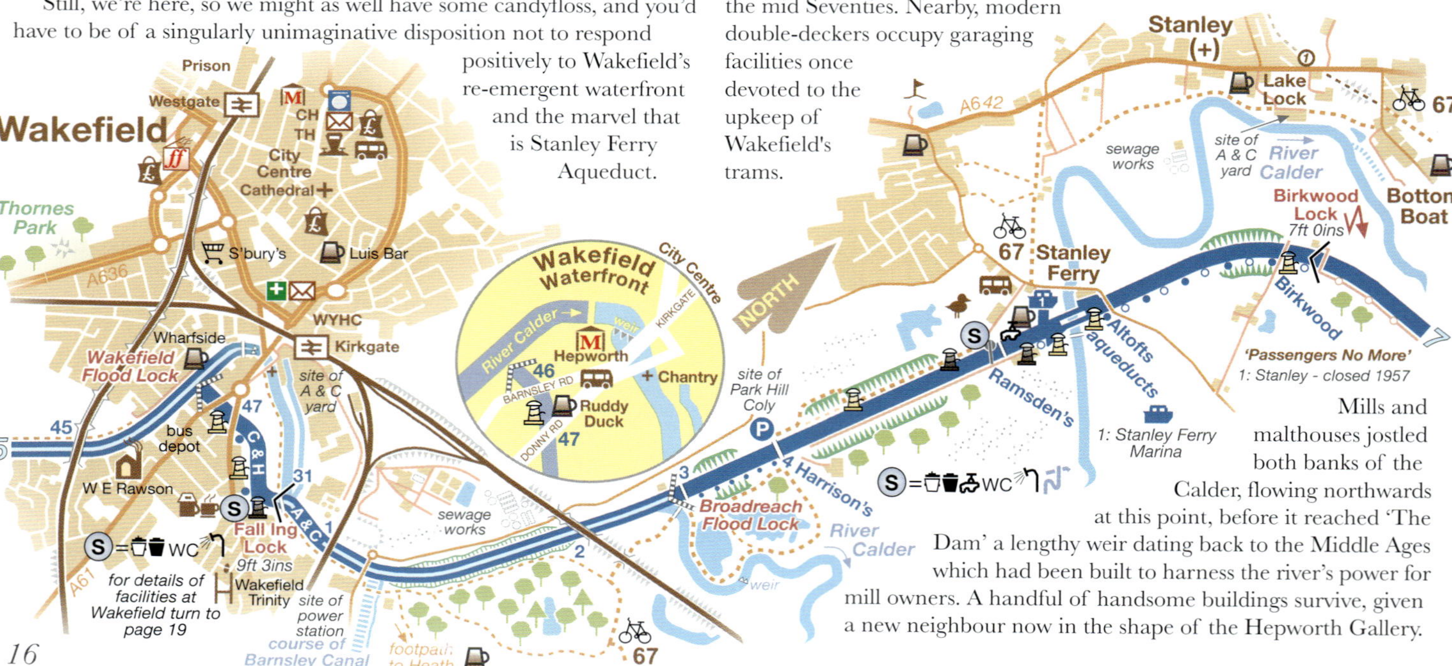

Mills and malthouses jostled both banks of the Calder, flowing northwards at this point, before it reached 'The Dam' a lengthy weir dating back to the Middle Ages which had been built to harness the river's power for mill owners. A handful of handsome buildings survive, given a new neighbour now in the shape of the Hepworth Gallery.

The Flood Lock lies at right-angles to the main channel. In the broad context of the river, steering into its embrace seems like an 'eye of the needle' exercise, especially if the current is strong. Mercifully, apartments have replaced British Waterways somewhat less than prepossessing 1950s corrugated iron clad warehouse. Visitor moorings are provided between the Barnsley (46) and Doncaster (47) road bridges with a new-build Marston's pub opposite. And you really should take time to explore Wakefield, for the old county town of the West Riding has some inspiring architecture, not least the adjacent bridge chantry chapel. One of a lovely quartet of such survivors - along with Rotherham, St Ives (Cambs) and Bradford-on-Avon - it is tempting to concoct an inland waterway odyssey encompassing them all.

Aire & Calder Navigation

The obdurate tail gates of Fall Ing Lock swing open to the new world of the Aire & Calder. A narrow, and now unnavigable channel parallels the Calder upsteam to the navigation's original terminus where the company's office (prior to removal to Leeds in 1850) remains. It is remarkable to consider that the Calder was rendered navigable to Wakefield at the very beginning of the 18th century.

The broad waters of the Calder sweep down beneath the euphoniously-named (though misspelt) Foundry Shoal railway bridge and past the levelled site of Wakefield Power Station, closed in 1991. Little evidence remains, either, of Heath Lock, entry point to the Barnsley Canal, abandoned in 1946, but the subject of long term restoration proposals, not least hereabouts, where a scheme to redevelop the power station site could incorporate a revived length of canal. Eleven miles long with fifteen broad locks, the

Barnsley Canal linked up with the Dearne & Dove Canal to provide a through route to the Sheffield & South Yorkshire at Swinton near Rotherham.

A stone arched railway bridge, conveying George Stephenson's Manchester & Leeds Railway of 1840 across the navigation, frames the approach to Broadreach Flood Lock and the commencement of the Calder Cut, scarcely a year senior to the railway: what upheavals in the landscape the mid-19th century must have witnessed; and all with spades and wheelbarrows.

By Harrison's Bridge (4) was the loading staithe for Park Hill Colliery. Hargreaves motor barges loaded coal for Ferrybridge Power Station here until 1982. Ten years later it was the site for the IWA's National Rally. Over the cutting's swarthy rims nature reserves are swifty reclaiming opencast coal workings. Long lines of moored boats lead to Ramsden's Swing-bridge, beyond which most boater requirements are catered for.

Nineteenth and twentieth century aesthetics step into the boxing ring at Stanley Ferry and there's a knockout in the first round. No prizes for guessing the winner. It's a mismatch between the original aqueduct of 1839 and its 1981 successor. Designed by George Leather - whose work you will encounter again as you explore West Yorkshire's waterways - it consists of an iron trough suspended between bowstring girders with classical ornamentations: an elopement of 19th century England with Ancient Greece which is said to have spawned the Sydney Harbour Bridge. British Waterways claimed that the old bridge wasn't up to taking the strain of modern commercial traffic, and elected to retire it, by-passing it with a pre-cast concrete trough. Within a decade there was no

continued overleaf:

continued from page 17:

commercial traffic to speak of. Now, happily, both structures are in use. There's little doubt that the original bridge is more fun to cross, but the characterless modern upstart beside it at least offers a better view of its predecessor's startling ornamentation.

Stanley Ferry's other claim to waterways fame is that it is the location of the Canal & River Trust's lockgate manufacturing workshops, one of only two such establishments still in production, the other being at Bradley on a remote backwater of the Birmingham Canal Navigations. For many years, these premises at Stanley Ferry were employed in the maintenance of the fabled 'Tom Puddings', the tug-hauled compartment pans invented by William Hamond Bartholomew (1831-1919), the A&C's most illustrious engineer and manager, for carrying coal. One of the 'Puddings' regular sources of the black stuff was St John's Colliery, whose basin lay opposite the yard. At one time - as archive photographs vividly testify - the pans were hoisted individually out of the water and conveyed along a mineral railway by steam shunting engines for loading direct at the pit head.

Accompanied by a well-surfaced towpath - which is designated part of the Wakefield Wheel, a forty mile circular route around the city - the navigation proceeds briskly through cuttings to Birkwood Lock, first of the push-button brigade if you're journeying east: previous experience preferred but not essential. The Calder meanders sinuously to the north, an erratic course which used to be pinpointed by the prominent twin lantern towers of St Peter's church at Stanley to the west. It was a landmark we always looked out for, and assumed we'd lost our bearings on our most recent research trip. The church, it transpired, had been demolished on safety grounds. Built in 1822, with Napoleonic War reparation funding, it had survived being gutted by fire in 1911, but not the indifference of the Secretary of State for Communities & Local Government (a Yorkshire-man, at that!) to this unusually beautiful building's fate. Not all the most anti-social vandals live on benefits and wear hoodies. To add insult to injury, St Peter's is/was the last resting place of William Hamond Bartholomew, following in his father's footsteps, as the A&C's engineer. The family lived at the company's maintenance yard at Lake Lock, its location on the riverbank predating opening of the Calder Cut. Many of its buildings remain in use as housing: the sort that values its privacy. In later years the yard was accessed via Foxholes Lock (Map 7). The world's first 'public railroad' was opened at Lake Lock in 1798. Formerly a rough and tumble mining community, Bottom Boat - which also boasted a coal staithe - is now home to the UK's leading footwear distributor. Determined to pay homage to Bartholomew, we searched fruitlessly for his grave in St Peter's churchyard, concluding that his stone must have been a casualty of the demolition, subsequently learning that it had been 'saved'. RIP wherever you are WHB, resident genius of the A&C.

continued from page 20:

Stephenson's Bridge (No.8), one of three arched railway bridges spanning the river and erected almost organically from local magnesian limestone, derives its name from the fact that this was George Stephenson's design for George Hudson's York & North Midland Railway; Hudson being the 'Railway King', an upstart draper determined to 'mak all railways cum t' York', and his fortune in the process. A hundred and seventy-five years later trains still rumble over it, giving you the opportunity to wave at poor call centre drones on their way home to microwaved suppers in Castleford, Glasshoughton and Pontefract, and similar apotheoses of Western Civilisation.

Bowdlerized barges lie moored by Methley Bridge. Floodbanks limit the boaters' view to 'Cas', hunched over the horizon like someone slumped over a bar. Power lines festoon the fields, glassily reflected in isolated oxbow meanders of the Calder's old course. One of these, variously known as 'Whitwood Mere' or 'Pottery River', remained navigable for many years to reach a bottle factory, brewery and pottery works. In the rheumy distance beyond Allerton Bywater the gracious facade of Ledston Hall reminds one that there was a degree of pre-industrial grace in the vicinity. Nay dear, this is not a nadir, this is just another Slough of Despond.

Wakefield Map 6

From a distance, Wakefield's skyline looks rather alluring, what with All Saints' soaring spire, and the prominent towers of both the town and county halls; for this, not Leeds, was the administrative seat of the West Riding. Reality is rather less encouraging. Garrotted by ring roads, and scarred by an acne-like rash of Poulsonesque Sixties architecture, Wakefield reminds you of Dr Johnson's adage that some places are 'worth seeing, but not worth going to see'. And yet, and yet, there are some absolutely wonderful buildings in Wakefield which are owed a better setting - the aforementioned Cathedral, Town Hall and County Hall; the lovely Chantry Chapel, and the revitalised (if seemingly uninhabited) Kirkgate railway station - and each time we revisit this city our admiration and affection increase exponentially.

Eating & Drinking

CREATE - Wakefield One, Burton Street. Tel: 01924 332330. Colourfully furnished cafe adjoining library and museum open 10am-4pm Tue-Sat. WF1 2EB
DAMELIO - Wood Street. Tel: 01924 674159. Stylish Italian wine bar. Open lunch and evening (from 5pm) Tue-Thur and from noon Fri-Sun. WF1 2ED
GYROS BROS - Market Street. Tel: 01924 682117. Greek restaurant from noon daily. WF1 1DH
LUIS BAR AT FERNANDES BREWERY - Avison Yard, Kirkgate. Tel: 01924 386348. Real ale cornucopia belonging to Ossett Brewery. WF1 1UA
IRIS - Bull Ring. Tel: 01924 367683. Highly regarded contemporary British restaurant. WF11HA
KING'S ARMS - Heath Common. Tel: 01924 377527. Classic stone-floored, gas-lit country inn. Bar & restaurant meals. Ossett Brewery ales. WF1 5SL
MARMALADE - Bull Ring. Tel: 01924 200203. Pleasant tea room open 9am-5pm, Mon-Sat: scones, cakes, pastries and soups. WF1 1HB
RUDDY DUCK - Bridge Street. Tel: 01924 379079. Marston's newbuild opposite visitor moorings with emphasis on pizzas. Open 10am daily. WF1 5JR
THE WHARFSIDE - Thornes Lane Wharf. Tel: 01924 360772. Refurbished pub overlooking the river opposite the Flood Lock. Mooring rings and ladder on river wall for boating patrons. WF1 5RL
WETHERBY WHALER - Calder Island. Tel: 01924 298800. Slick fish & chip restaurant and takeaway adjacent Calder & Hebble Navigation by Thornes Lock (Map 5). From 11.30am (noon Sun). WF2 7AW

Shopping

The centre of Wakefield is ten minutes walk uphill from the river, though enterprisingly a 'free city bus' operates at ten minute intervals from Stop 4A on Barnsley Road opposite the Ruddy Duck. Trinity Walk and The Ridings are focusses of retail therapy, but naturally there are more intriguing independent outlets to be discovered by discerning shoppers. Both Allums (Brook Street) and Hofmann's (various outlets) are keen to stress the prize-winning credentials of their pies and sausages; as befits the location the latter also do a rhubarb pie. The annual Wakefield Rhubarb Festival takes place in February.

Things to Do

THE HEPWORTH - The Waterfront. Tel: 01924 247360. Marvellous gallery celebrating the sculpture of Barbara Hepworth who was raised in the city. Also features Castleford's world famous Henry Moore and less famous Albert Wainwright. Shop, cafe. WF1 5AW
CHANTRY CHAPEL - Wakefield Bridge. Tel: 01924 373923. Services on 1st and 3rd Sundays at 4.30pm. Also open to view on selected dates under the auspices of Wakefield Historical Society. WF1 5PL
CATHEDRAL - Northgate. Tel: 01924 373923. Holds claim to the highest spire (247ft|) in Yorkshire. Earliest parts date from 15c. Shop and cafe. WF1 1HG
MUSEUM - Burton Street. Tel: 01924 305356. New museum & library with cafe - see E&D. A medieval logboat - recovered from the Calder during excavations for Stanley Ferry Aqueduct in the 1820s - is on display in the library. WF1 2DD
GEORGE GISSING MUSEUM - Thompsons Yard off Westgate. Tel: 01924 255047. Childhood home of unfairly forgotten Victorian novelist. Open Saturday afternoons 2-4pm May-Sep. WF1 1XH
SANDAL CASTLE - Marygates Lane. Tel: 01924 249779. Grounds open daily, visitor centre Wed-Sun, 11am-3pm. Catch bus 106 from stop near Barnsley Road visitor moorings. WF2 7DS
WEST YORKSHIRE HISTORY CENTRE - Kirkgate. Tel: 0113 535 0142. Archive centre. WF1 1JG

Connections

BUSES - services 96 and 108 run to Yorkshire Sculpture Park at West Bretton. Tel: 0871 200 2233.
TRAINS - stations at Westgate and Kirkgate: the former more important, with InterCity services to Leeds, London and the Midlands; the latter more handily placed for the river, and offering useful links for would be towpath walkers with Castleford and Knottingley. Tel: 0345 748 4950.
TAXIS - Wakefield Cars. Tel: 01924 380000.

Stanley Ferry Map 6
Eating & Drinking

THE STANLEY FERRY - Ferry Lane. Tel: 01924 290596. Waterside all-day family-orientated pub with 'Wacky Warehouse'. Good value food. WF3 4LT

Connections

BUSES - services 125, 146/7 provide approx 20 minute interval connections Mon-Sat with Wakefield, and Castleford. Tel: 0871 200 2233.

7 AIRE & CALDER Altofts & Methley 4mls/2lks/1½hrs

FROM Altofts Ings to Methley Mires the Aire & Calder's Wakefield branch purposefully thrusts its way across an area once riddled with coal mines but now reverting to pre-industrial verdure.

Sometimes, however, you can't help thinking that nature needs constant supervision. Unchecked, all those burgeoning saplings will grow into the sort of primeval forest the Romans must have encountered marching north to bring pizza and pasta to the Brigantes. Unmanaged woodland is like unshaven armpits - not a pretty sight.

Off-stage the 21st century world of motorways, business parks and cold storage facilities is going about its business, but down on the water we're left to our own devices; daydream believers and home-coming queens. When Pearson's first charted these waters in the mid-Nineties every lock had an avuncular keeper in its control cabin. But trade was evaporating and their days were numbered - you could see DOUBT stitched on their leathery foreheads. We valued their *bonhomie*, their *savoir faire*, and the way they called us 'cock' with an easy familiarity, as though we were about go whippet-racing together. Push-button panels are a pale substitute; like a vending machine in place of a buxom barista.

But Progress goose-steps obliviously on as it has always done. Everywhere you look the past tense applies. The lane which gives its name to Kings Road Lock led, apparently, to a ferry across the Calder and thence over the ings to Temple Newsam (Map 13). Foxholes Lock - closed circa 1962 - had four pairs of gates, so that it could be negotiated whether the river level was above or below that of the canal. Altofts and Fairies Hill locks were abandoned in the 1950s, though the latter has been reinstated to provide off-river moorings. But the most recent change of tense occurred at Whitwood Wharf, commissioned in 2002 to receive aggregates from quarries alongside the River Trent in Nottinghamshire. Lafarge's laudable use of water transport was keeping an estimated twenty-five thousand lorry movements off the roads annually until they were taken over by Tarmac in 2013 and the use of barges ceased abruptly.

continued back on page 18:

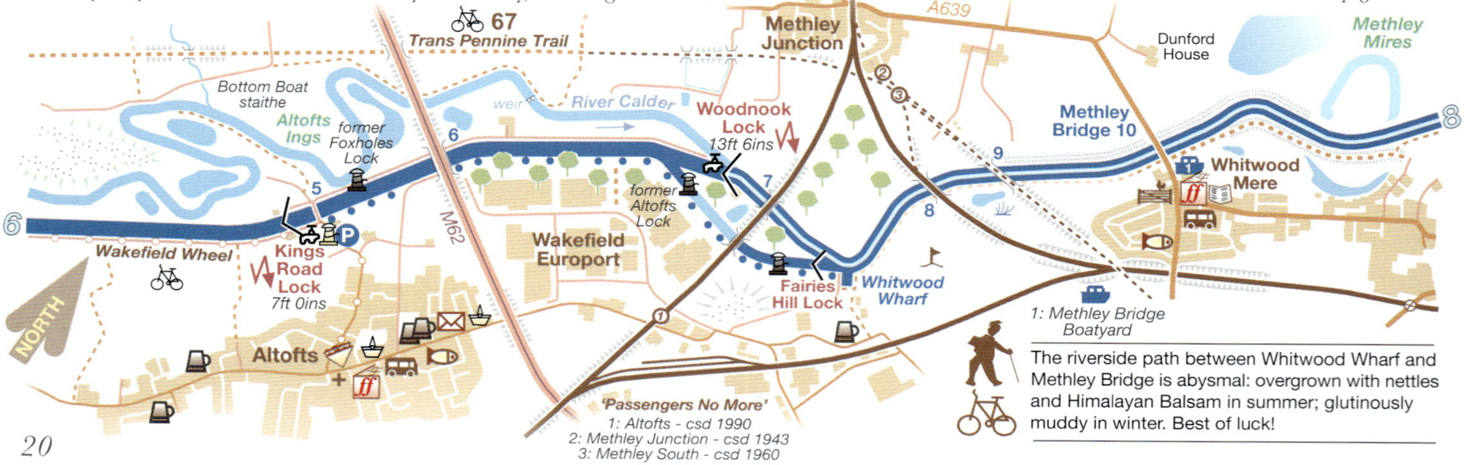

The riverside path between Whitwood Wharf and Methley Bridge is abysmal: overgrown with nettles and Himalayan Balsam in summer; glutinously muddy in winter. Best of luck!

DID you know 'Cas' when it hooched with barges; push-tows and compartment 'trains': the oil, the effluent, the coal? If you didn't, then you have our condolences, for you have missed out on a way of life and richness of scene which no amount of leisure boating can come close to replicating. And it's not only the commercial craft that are absent, it's the characters whose life's-work was to operate them:

comparatively empty waters and denuded quasi-industrial landscapes an inescapable sense of post Bastille Day *tristesse, un couverture humide* from which only a compelling guide can effect convincing and lasting psychological escape. Fortunately, that time-honoured variety act of J. M. Pearson & Son are on hand to conjure coloured doves from feral pigeons.

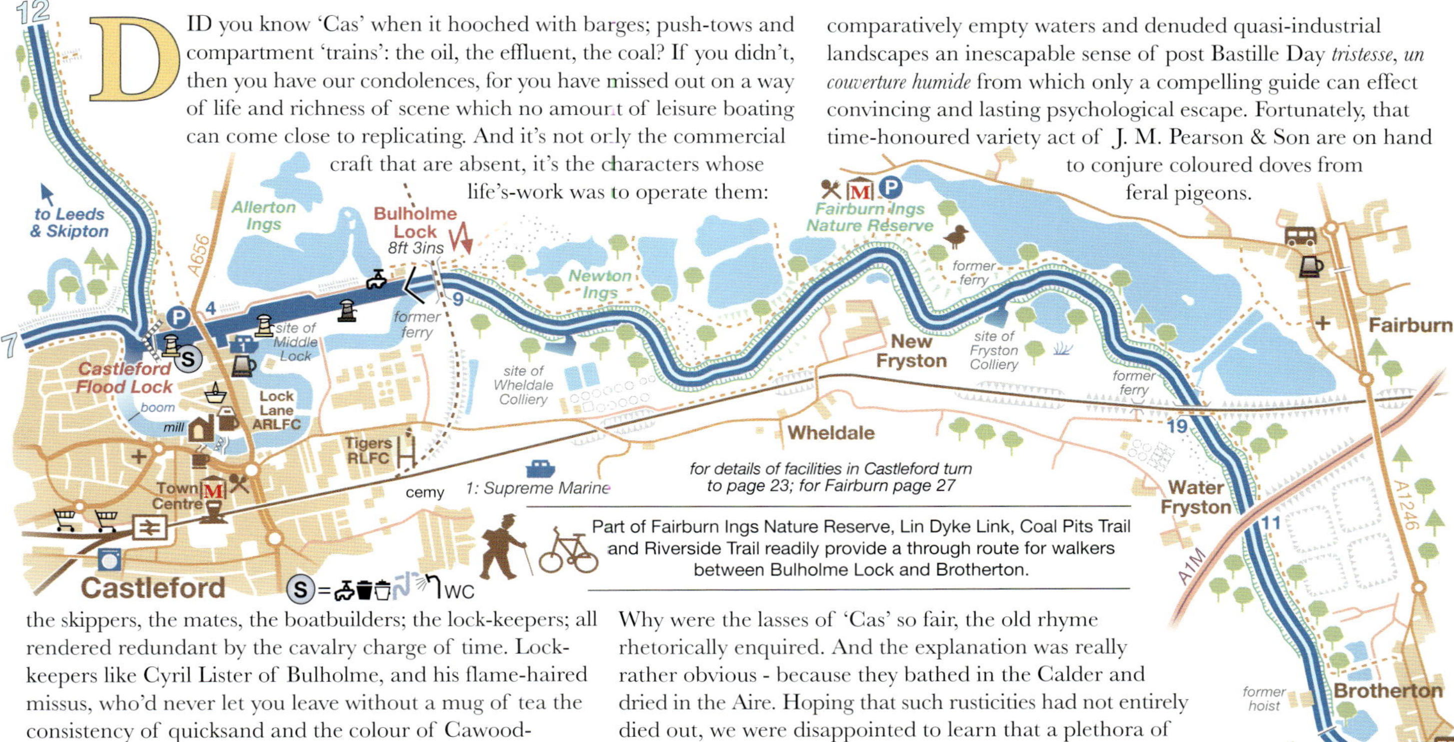

the skippers, the mates, the boatbuilders; the lock-keepers; all rendered redundant by the cavalry charge of time. Lock-keepers like Cyril Lister of Bulholme, and his flame-haired missus, who'd never let you leave without a mug of tea the consistency of quicksand and the colour of Cawood-Hargreaves keels. Why did they go? Through no fault of their own. But, rather, because the industries they served have vanished too: the coal mines (deep and opencast), the chemical works, the oil depots, the potteries, the flour mills, the glass works, the aggregates wharves. Thus, there hangs over these

Why were the lasses of 'Cas' so fair, the old rhyme rhetorically enquired. And the explanation was really rather obvious - because they bathed in the Calder and dried in the Aire. Hoping that such rusticities had not entirely died out, we were disappointed to learn that a plethora of tanning studios in the town had rendered the tradition all but obsolete. We sought solace in the confluence of the two rivers - a counterfeit aquatic crossroads - but lighted upon a burgeoning housing development: did no one warn them of the rivers' propensity to flood? *continued overleaf:*

continued from page 21:

The combined waters of the Aire and Calder go tumbling over Castleford Weir, 'Cas's' not entirely unconvincing answer to Niagra Falls. A millennium inspired footbridge spans the boiling waters beside a retired flour mill. The bow of a sunken barge rears out of the water like an art installation. Thoughtfully, the authorities provide a boom upstream to ensure that life doesn't imitate art. The modern-day boater, however, sees nothing of this as they press on through Castleford Cut, opened in 1775 to upgrade the navigation which previously simply by-passed the weir. Access to the river downstream thereafter was via Castleford Middle Lock until Bulholme Lock extended the cut in 1828. Like Pearsons, the A&C had a policy of continual improvement.

Shrugging off the Flood Lock's enthusiastic embrace, spare time to take stock of your surroundings, noting that there were once *two* entry locks to the cut, creating a small island, now solely occupied by the remaining lock's *Marie Celeste* of a control cabin. In Lock-wheeler's salad days as a cub reporter on *Waterways World*, he'd pop his head in here to ask if anything commercial was moving, and was rarely disappointed. Back then, British Waterways had a traffic office adjoining the flood lock, a risible concept now. Once there were cottages on the island. Now just benches paying homage to watermen long gone to a waterway still thronged with keels and tugs and Tom Puddings in the sky.

3-day visitor moorings and boater services are provided between the flood lock and Bridge 4; further accommodation being provided through the bridge on the offside between what used to be Cawood-Hargreaves boat dock (but is now Supreme Marine) and the site of Castleford Middle Lock, which finally fell out of use during the Second World War. A coal-loading shute juts melancholically out from the bank at the site of the former lock, smokeless fuel was discharged from lorries here.

High, wide and handsome, Castleford Cut strides steadfastly down to Bulholme Lock. Chemical works between the cut and the river have been cleared pending housing developments. Fitted with intermediate gates - within which leisure craft appear tiny today - Bulholme's 460ft chamber was long enough to pen a train of 19 Tom Puddings. Moreover, additional craft could pass through simultaneously by taking refuge in the widening provided at the top end of the lock. The absence of such activity makes Bulholme very quiet now; we even heard a curlew calling over the ings above the subterranean groans of the electrified sluices. See how the lock-keeper's bungalows were erected on stilts, given the Aire's propensity to flood. An impressive, albeit decrepit railway bridge spans the river below the lock. It once carried the Castleford to Garforth branch line, closed to passenger trains way back in 1951, but used by coal trains from Allerton Bywater Colliery until its closure in 1992. Some of the trackwork remains in place and there have been vague proposals to re-open the line as part of West Yorkshire's commuter network.

A sinuous river section - colloquially known to generations of bargemen as the 'five mile pond' - separates Castleford from Ferrybridge. Historically, land paralleling the north bank consisted of low-lying ings, an old Norse word, prevalent in Yorkshire, describing land prone to flooding. Years of colliery waste dumping, however, followed by subsidence, has resulted in a fractured landscape, increasingly clothed in trees. Certainly old boatmen would fail to recognize it. It could be the Congo River. You could be Joseph Conrad's Marlow, trading ivory with cannibal tribes and accompanied by an altogether different kind of pilgrim to those you set out with from Sowerby Bridge. Cormorants and divers emphasise the Aire's transformation from commercial highway to natural watercourse.

The basins at Wheldale and Fryston closed, along with their respective collieries, in the 1980s. Son of a Castleford miner, the sculptor, Henry Moore, was commissioned to draw scenes underground at Wheldale Colliery during the Second World War. So the coal basins survive, but there is hardly any clue at all to the former arms - such as Staniland's - which penetrated into quarry faces along the magnesian limestone ridge upon which Fairburn stands. Ferries plied to and fro across the Aire - as much to carry boat horses as people - but they are long gone. Fairburn Ings is an extensive RSPB wetland habitat developed from former mining subsidence. It attracts all manner of wildlife - and humans too - but no one has yet had the gumption to provide pontoon moorings.

George Stephenson's brick-built, tie-bar strengthened railway bridge (renumbered 19 by CRT for reasons best known to themselves) is another of his structures for the Y&NMR. A useful footbridge, bolted onto the southern side, fulfills the role of a long lost ferry. At intervals, goods trains rumble cacophonously across it, though none carry coal anymore, as was predominantly the case when this guide was first published in 2015.

Ferrybridge 'C' Power Station closed in 2016, the last of three coal-fired plants which had been generating electricity on the site since 1927, much of their fuel being brought in by barge. In an echo of the old Tom Pudding trains, a fleet of purpose-built tugs and 'pans' - three pans in each set with a payload of 500 tonnes - represented the river's final manifestation of water transport. Failures of the tipping machinery - which lifted each pan bodily out of the water and dropped its contents onto conveyor belts - and wear and tear on the pans rendered the whole operation unviable. The last delivery of coal by water occurred on 17th December 2002, but the hoist is still there, part of an alien assemblage of decaying apparatus, lining the river like an abandoned space station. Many of the obsolete pans, we learned, went south to be employed as the hulls of Thames houseboats, an altogether more genteel existence. A new multi-fuel, waste to energy plant sustains Ferrybridge's generating role.

Like the A&C before it, the A1(M) has been the subject of a rolling programme of improvements. Lagentium Viaduct (Bridge 11) carries a section of road opened in 2006 across the Aire. Lagentium was the Roman name for Castleford. The village of Brotherton sits demurely beside the old Great North Road, and the sooty tower of St Edward the Confessor peeps over the rooftops of enviable riverside properties. Gravestones in the churchyard hint at high rates of infant mortality in the late 18th and early 19th centuries. An engraved stone on the downstream abutment of the railway bridge records that as originally built in 1850, to the design of Robert Stephenson, it was a tubular bridge, similar in style to his better known structures across the River Conway and Menai Strait. Coincidentally, he was assisted in those projects by a William Fairburn, not of this parish!

Castleford

Quite the best way to approach this inimitable Rugby League town from the visitor moorings (either side of Bridge 4) is to follow Lock Lane as far as Mill Lane (just before the petrol station) and then to make a zig-zag right and left so as to cross the Aire on the impressive millennium footbridge which snakes its way across the river's foaming weir. On reaching the town centre it is hard to see any evidence of the fact that this was once a Roman settlement, and as the area's old heavy industries continue to contract the local economy relies increasingly on the manufacture of exclusive Burberry garments, a paradox which won't be lost on Canal Companion regulars.

Eating & Drinking
THE GRIFFIN - Lock Lane. Tel: 01977 324435. Pictures of old barges and rugby league teams. Open Tue-Fri from 2pm and from noon at weekends. WF10 2LB

JOY BANG_A - Bridge Street. Tel: 01977 524293. Popular Indian restaurant and take-away. WF10 1JS
NO. 20 - Wesley Street. Tel: 01977 667116. Stylish (for 'Cas'!) cafe bar and grill. Open daily from 8.30am. Evenings Fri & Sat. WF10 1AE
QUEEN'S MILL TEA ROOMS - Aire Street. Tel: 0781 010 2116. Open 10am-4pm (11am Sun). WF10 1JL

Shopping
The Carlton Lanes shopping precinct hosts many well known retailers (and also, note, an outlet for Castleford Tigers merchandise) but one warms more towards the extensive market hall open daily Mon-Sat (half-day Wed), and its attendant outdoor equivalent held on Mon, Thur, Fri & Sat. If you're rushing through, then the little convenience store on Lock Lane stocks a range of food and drink and does a nice line in sustaining hot take-aways. Beer lovers will want to beat a path to The Ale Seller on Aire Street.

Things to Do
CASTLEFORD FORUM MUSEUM - Carlton Street. Tel: 01977 722084. Romans, coal mining and rugby league are among the themes of this small museum above the library. Some nice period photographs by local man Jack Hulme. In the Local Studies room on the first floor is material relating to the town's favourite son, the sculptor Henry Moore. Open daily ex Wed & Sun from 9.30am. WF10 1BB
QUEEN'S MILL - Aire Street. Tel: 01977 556741. Iconic riverside stoneground flour mill under the care of Castleford Heritage Trust. Visit internet for up to date details regarding open days. WF10 1JL

Connections
TRAINS - Northern services to/from Leeds, Wakefield (for rhubarb), Knottingley (for snails) and Pontefract (for liquorice)! Tel: 0345 748 4950.
TAXIS - Arrow Cars. Tel: 01977 553333.

HERESY, in some quarters, but we miss Ferrybridge's cooling towers. The Aire's broad sweep down to John Carr's handsome three-arched bridge of 1804 seems bereft without their brooding presence. Carr, previously encountered at Horbury (Map 5) and arguably best known for Harewood House near Leeds and Buxton Crescent, was a prolific bridge builder, and presumably responsible for the adjoining toll house as well, now in use as a coffee shop. Good visitor moorings can be found offside after the flood lock, or at the entrance to the river section itself. Overlooked by a flour mill and a pair of bottle factories,

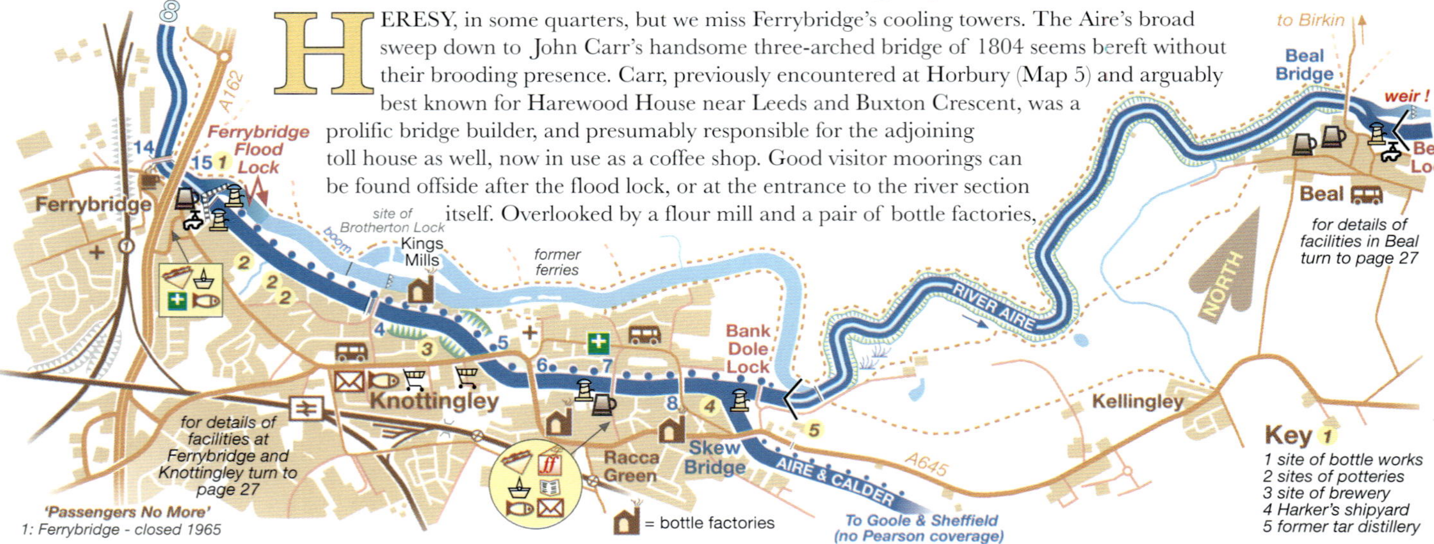

the navigation essays a mildly serpentine course through Knottingley.

Beyond Shepherds Bridge (No.8) lay Harkers renowned shipbuilding yard, and a motley assembly of craft still gather here. The main line veers to the right (or starboard in this maritime atmosphere) and makes purposefully off for Goole, a route, alas, not presently interpreted by Pearsons. Meanwhile, east of Knottingley, Selby-bound pilgrims lock down through Bank Dole Lock onto the River Aire. Dole by name and somewhat doleful by nature - an atmosphere exaggerated by its derelict lock-keeper's house - the chamber boasts pairs of upper gates facing in either direction, though it's difficult to imagine anyone navigating safely off the Aire if it was higher than the canal. But there's no disputing the adrenalin surge you experience once the heavy bottom gates have been opened and you move out onto the river; why, it feels like floating in mid Aire.

Meandering extravagantly from the outset, the river makes its way towards its confluence with the Ouse at Asselby Island, half a dozen miles upstream from Goole. It all seems a far cry from Gargrave (Map 21) let alone its source at Malham up in the Yorkshire Dales. Quite literally, the river boxes the compass: one minute the sun is burning your neck; the next glaring in your eyes.

Beal Bridge was once an ancient timber affair with a toll payable to cross it. Passage is free now, but the replacement span is correspondingly dull. Flotsam and jetsam, left high and dry, on its piers only serves to underline the river's propensity to flood. Charming visitor moorings (plus a water tap and nearby pub) are available in the lock cut.

10 RIVER AIRE and SELBY CANAL   West Haddlesey   5mls/0lks/1½hrs

TUNNEL vision is not without its rewards. The Aire's high banks focus attention on the route ahead, and its corkscrew character ensures the view is constantly changing. Thus there are not *three* power stations, just poor abandoned Eggborough seen from different angles. And if cooling towers leave you underwhelmed, surely you can't fail to be captivated by Kellington's isolated parish church of St Edmund's which crops enchantingly up on a couple of occasions. Two theories vie for explanation of the church's remote setting: to lessen the odds on flooding, or to be at the epicentre of a parish of five villages. In the early Nineties, British Coal paid to have the tower dismantled, stone by stone and re-erected, because a new seam at Kellingley Colliery threatened to destabilise its already fragile fabric.

Entering through sluices when flood prevention demands, Old Eye represents the original course of the Aire, all but forming an island now of rich arable farmland. In *Yorkshire's River Aire* (Terence Dalton 1976) John Ogden reveals that the river hereabouts is flowing through a deep trench of thick clay soils formed during the Pleistocene Ice Age. Apt to lose concentration when Geography O Level turned geological, we must take his word for that, but this still reasonably easily acquired secondhand book is worth adding to your shelves for the evocative painting of a barge - tranquilly moored to a high-banked, willow-fringed bend in the river - which adorns the dust-wrapper. Forty years on, Ogden's letterpress printed text and half-tone illustrations provide an almost historical commentary on the Aire's journey from Malham to the Ouse, but at Haddlesey Flood Lock boaters must take their leave of the river - which still has fifteen unnavigable miles to flow before it reaches its confluence with the Ouse - and join the Selby Canal.

A change in environment becomes immediately apparent. Not because the canal doesn't do its best to impersonate a river, but because the water beneath your hull is a stolid and weedy soup, prone to duckweed in summer.

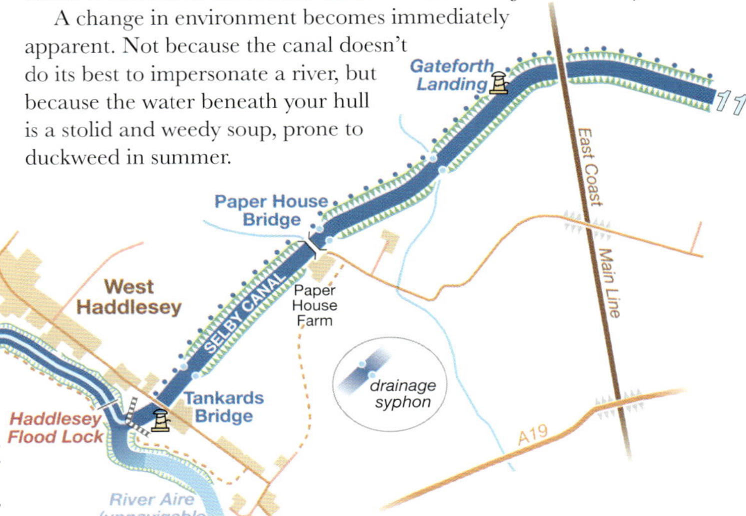

ALONG with the Ripon Canal and the Pocklington Canal, the Selby Canal belongs to an alluring trio of semi-detached canals in Yorkshire. Semi-detached, that is, in the sense that all three are not connected to other canals, but to various tributaries of the Ouse. An experienced boat crew might derive much enjoyment from 'joining the dots', though to do so would involve negotiating potentially tricky tidal waters, hence the emphasis on *experience*.

Opened in 1778 to by-pass the tidal Aire, the Selby Canal's initial success was short-lived as first the Knottingley & Goole Canal and then the Leeds & Selby Railway ate into its profits. It remained, however, a useful link between the West Riding and York, its dwindling trade not petering out entirely until the 1970s. These days it's a soporific backwater, a route for canal cognoscenti, purists and pilgrims, its shallow nature leaving no alternative but to throttle back and enjoy the ride.

To the north, Brayton Barff stands out, a curious hundred and fifty foot high pimple on an otherwise flat plain. The canal prods a rural finger deep into Selby's soft urban underbelly, so that it is really only the last half mile or so that is compromised by works and retail parks and housing. Redbrick abutments mark the course of a chord off the former Selby & Goole Railway which enabled goods from the West Riding to reach Goole and Hull without interrupting shipping by opening the swing-bridge at Selby. William Jessop engineered the Selby Canal, but you can be safely assured he was not responsible for the bridge carrying the A63 across the canal, even if it bears his name. Passing a boatyard - welcome in a region not overly endowed with such facilities - the canal approaches its terminus, though first an electrically operated swing-bridge on a not un-busy road has to be negotiated; carry your CRT facilities

**1 lock if proceeding onto tidal Ouse

(Yale) key and your diplomacy with you. Thence the canal widens into a basin (where you can turn) with permanent moorings to the right and visitor moorings, parallel to the towpath - overlooked by balconied apartments - on the left. Mechanised Selby Lock may be, but passage down onto the Ouse is restricted to flood tide under the watchful eye of the lock-keeper, an approachable mine of local information.

Fairburn — Map 8

By-passed village on what was the Great North Road.

Things to Do

FAIRBURN INGS - Newton Lane. Tel: 01977 628191. RSPB nature reserve with trails and hides. Star species include green sandpipers, kingfishers, reed warblers and (once in a blue moon) bitterns. Shop, binocular hire, and refreshments; but *no* moorings! WF10 2BH

Ferrybridge — Map 9

Once a nodal point on journeys north by road, rail or water, Ferrybridge remains a centre for the generation of power, though naturally by means other than that discredited outcast, Old King Coal.

Eating & Drinking

Two 'locals', two sandwich bars, and one fish & chips. Coffees and light bites served (Fri & Sat 11am-3pm) at the old toll house by the eponymous bridge.

Shopping

Parade of shops on the High Street easily reached from the visitor moorings adjacent the Flood Lock.

Connections

BUSES - services 157 to/from Castleford; 148/9 to/from Knottingley, Pontefract and Wakefield.

Knottingley — Map 9

The fact that Knottingley was a medieval port is echoed by the number of seafarers buried in St Botolph's churchyard, together with the fatalities on the war memorial who served in the mercantile marine. Why even the illustrious Shipwrecked Fishermen & Mariners Society had premises on Aire Street up until the 1930s.

Eating & Drinking

RACCA GREEN FISHERIES - Racca Green (south of Bridge 7). Tel: 01977 672385. Superb fish & chips in the shadow of the bottle factories. WF11 8AT

Shopping

Morrisons and Lidl supermarkets. Sainsbury's 'Local' at Racca Green. Pharmacy north of Bridge 7.

Connections

BUSES - Arriva 476 operates approx bi-hourly (ex Sun) to/from Selby and Pontefract. Tel: 0871 200 2233. TRAINS - hourly Northern services to/from Castleford, Wakefield & Leeds. Limited Sunday service. Skeleton service to/from Goole. Tel: 0345 748 4950. TAXIS - Knottingley Cars. Tel: 01977 672238.

Beal — Map 9

A dormitory village, as quiet as the grave during daytime, solely the potato farmers make any noise, their tractors and trailers forever crossing and recrossing the river bridge, which luckily for them no longer charges a toll. Stretch your legs to Birkin and its lovely, pale-complexioned, largely Norman church.

Eating & Drinking

KING'S ARMS - Marsh Lane. Tel: 01977 235351. Friendly family run pub adjacent the river bridge, previously known as the Hungry Fox. Open from 4pm weekdays and from noon weekends. DN14 0SL

Selby — Map 11

Selby basks in the glow of its abbey whose stonework appears to have the knack of absorbing light like a solar panel. Bereft of shipping the tidal Ouse ebbs and flows beneath the town's paired swing-bridges, the railway one being topped by a picturesque control cabin. On Ousegate, blue-gated and red-bricked, stands the original railway terminus of 1834. Henry I was born in Selby while his father, William the Conqueror, was busy 'subjugating the north'. Choral evensong takes place in the abbey on the first Sunday in the month. There are cholera graves in the churchyard.

Eating & Drinking

CAPRI - Abbey Place. Tel: 01757 706856. Charming little Italian beside the abbey. Open Wed-Sat lunch and evening (from 5pm). YO8 4PF
MISTER C's - Micklegate. Tel: 01757 701913. Fabulous eat-in or take-away fish & chips. YO8 4PU
THE DOGHOUSE - Tel: 01757 703163. Park Street. Craft beer bar featuring beers from the Little Black Dog brewery near Goole. A range of homemade pasteries and savouries complements beers, ciders, spirits and artisan coffees and teas. YO8 4PW
OLIVE BRANCH - Gowthorpe. Tel: 01757 428010. Friendly Turkish restaurant (see also Sowerby Bridge) at the western end of the town centre. YO8 4HE

Shopping

Monday is 'market day'. Morrisons, Sainsbury's and Tesco supermarkets. Delightfully old-fashioned department store called Wetherells. Don't miss Mollie Sharp's Cheese & Deli outlet on Finkle Street.

Connections

BUSES - useful Arriva service 476 to/from Pontefract via Beal and Knottingley. Tel: 0871 200 2233. TRAINS - services to/from Hull, Leeds, York, Doncaster and King's Cross. Tel: 0345 748 4950. TAXIS - Station Cars. Tel: 01757 702567.

RIVERS change course almost as frequently as politicians - well, perhaps not *that* often. In 1988 the Aire's north bank gave way and flooded the opencast coal workings at St Aidan's to a depth in excess of two hundred feet, forming a two hundred and fifty acre lake in the process. The remaining coal was deemed so valuable that British Coal - in one of its last pre-privatisation gestures - footed the £20 million (treble that in today's value) bill for a new navigable channel of some two miles length to be dug. The project was undertaken by McAlpine's, who approached the job as if they were building a motorway. Then simply added water. In the process, the locks at Kippax and Lemonroyd were combined into one huge new chamber at the latter location. Not without irony, the rebuilt navigation coincided with a downturn in commercial carrying by water. Indeed, years have passed without any at all, though currently, we're happy to report, sea-dredged sand is being conveyed from Albert Dock, Hull to Knostrop Depot, Leeds by barge: a sight for sore eyes; after all, this was what the A&C was built for!

A prominent landmark to the north is an opencast coal dragline nicknamed 'Oddball'. Built in Milwaukee in 1946, it has been preserved as a memorial to 'sunshine miners' by St Aidan's Trust. Top speed: less than half a mile an hour! Its former stamping ground has post-industrially morphed into a notable Royal Society for the Protection of Birds nature reserve, a habitat of reedbeds, wetlands, meadows and woodland. Amongst the residents are bitterns and Cetti's warblers.

The floral environs of Woodlesford Lock re-emphasise the sea change which has come over the area, and with an accompanying vineyard in the vicinity you could fool yourself that you were on the Canal du Midi ... for a minute or two. Wine may be produced locally, but beer no longer is. Bentley's Yorkshire Brewery bit the dust in 1984, having been acquired by those arch predators of smaller fry, Whitbread.

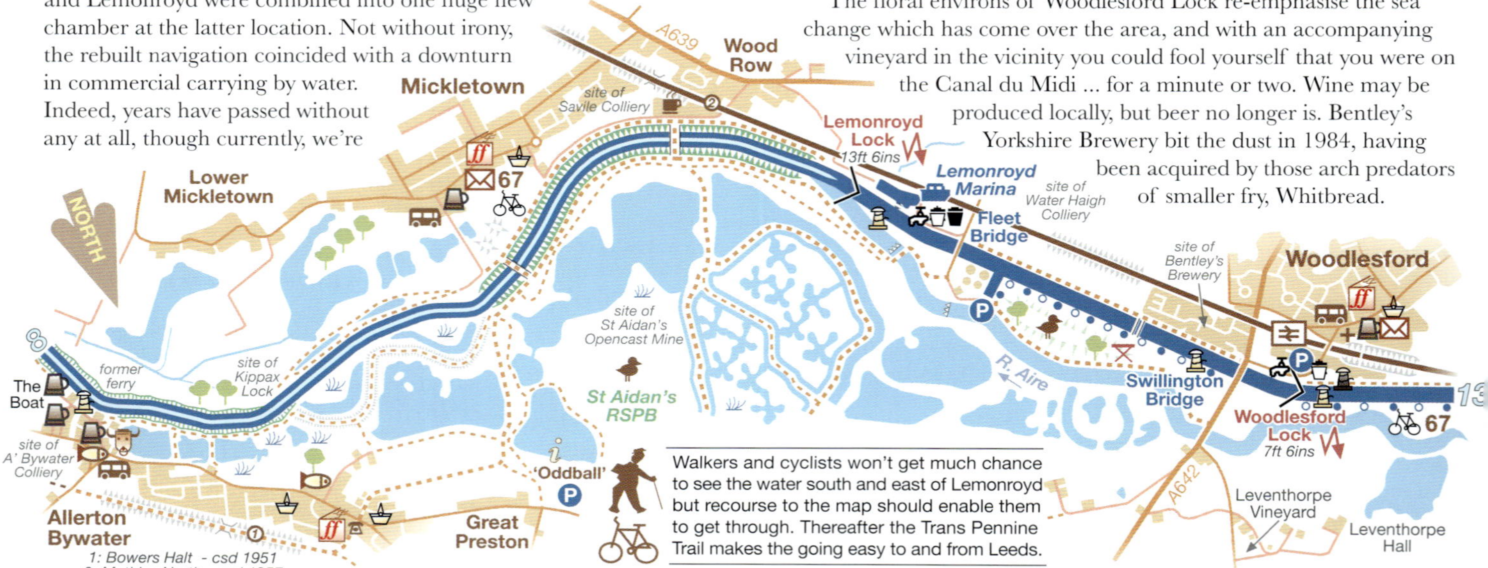

Walkers and cyclists won't get much chance to see the water south and east of Lemonroyd but recourse to the map should enable them to get through. Thereafter the Trans Pennine Trail makes the going easy to and from Leeds.

for summary of facilities at Allerton Bywater turn to page 34

for summary of facilities at Woodlesford turn to page 34

NORTHBOUND, you begin to sense Leeds before you see it; the high M1 motorway crossing at Bridge 7A acting as a sort of portcullis and harbinger of urbanisation. Across the compromised valley the Tudor-Jacobean facade of Temple Newsam does its best to ignore the motorway's cacophony. With grounds by Capability Brown and being often called 'the Hampton Court of the North', amidst its various claims to fame are that it was the birthplace in 1545 of Lord Darnley (husband of Mary, Queen of Scots) and the rural terminus (until 1959) of Leeds' tram route 22: somewhat less romantically, the number ten bus does the honours now.

At least Temple Newsam doesn't have to put up with Skelton Grange power station on its doorstep anymore. Its five cooling towers were demolished in 1998 following closure of the generating plant four years earlier. In the Sixties, Cawoods won the contract to deliver coal by water and had a fleet of 200-ton capacity motor barges built by Dunstons of Thorne specifically for the job. Though not all the coal came by water, concrete bowstring Bridge 7B carried a railway into the site.

The lay-by used by barges discharging coal is still evident, along with the concrete bollards they tethered to: mute testimony to vanished trade.

Thwaite Watermill closed as a commercial undertaking in 1976 in the wake of a devastating flood. It had been in the Horn family for over a hundred years, but could trace its history back to the middle of the 17th century when it was engaged in cloth finishing. Down the years Horns employed the mill to grind a variety of materials, many of them seemingly destined for use in the manufacture of commodities beginning with the letter P: pharmaceuticals, polish, pottery, paint and putty. Barytes was barged down the Leeds & Liverpool Canal from Cononley (Map 20); flint, collected along the south coast for use in the manufacture of china, was shipped to Goole and barged up the Aire & Calder. Evidence of all this industry might have vanished had not a society been formed to preserve the site, which is now operated as a museum by Leeds City Council. The mill and its associated buildings stand photogenically alongside the navigation. Furthermore, this is a good site to moor on the outskirts of the city, though you may be charged for the privilege.

continued on page 33:

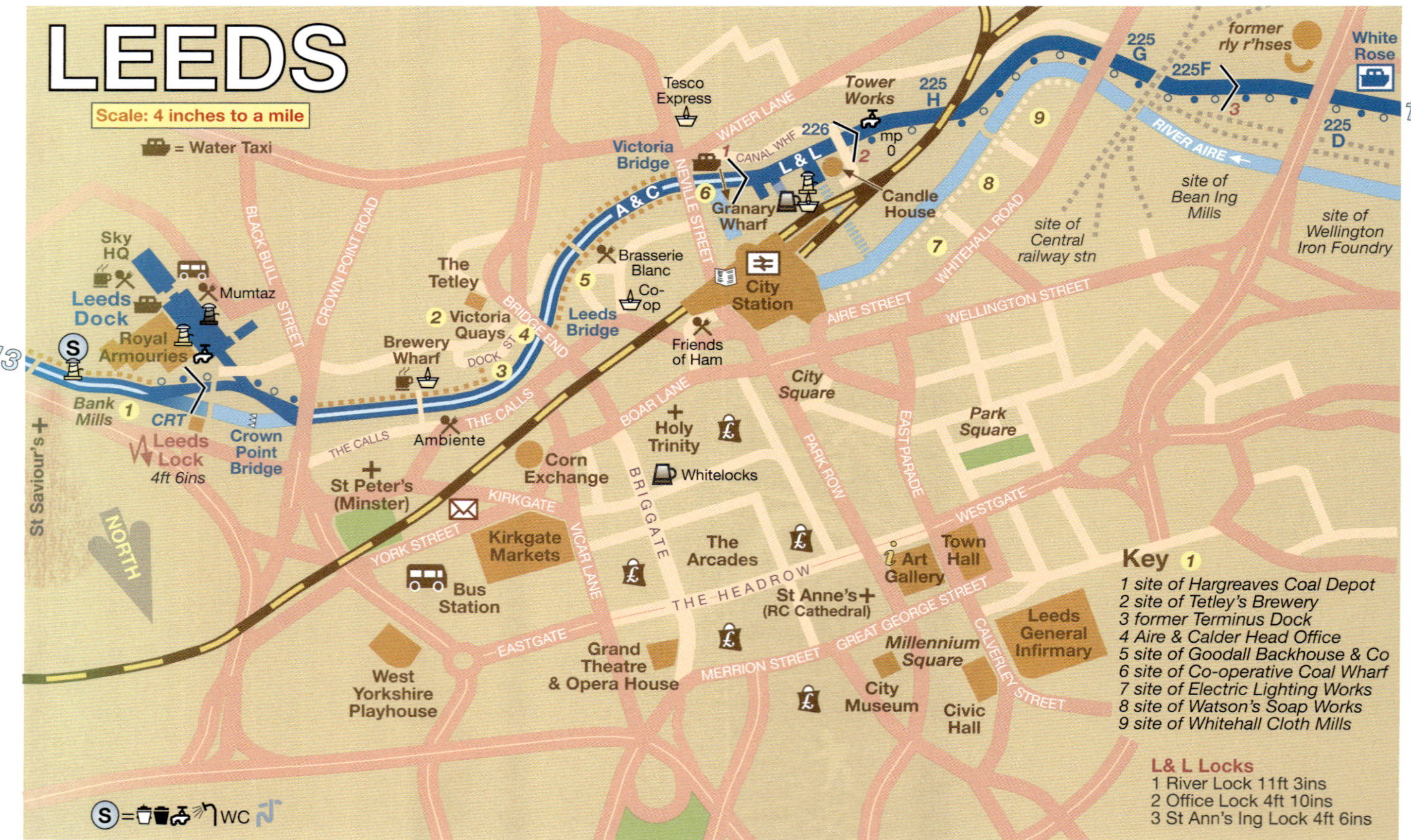

for details of facilities in Leeds turn to page 34

W E never tire of visiting Leeds, and quarter of a century after *Pennine Waters*, the progenitor of this guide, first appeared to a public barely prepared for its poetic outpourings, there was no gainsaying the pleasure to be back in Yorkshire's commercial capital, reacquainting ourselves with its waterfront for what was effectively the seventh time, professionally.

A mooring pontoon and boating facilities block are provided just downstream from Leeds Lock, the last on the Aire & Calder. Savour its push-buttons: Liverpool lies ninety-seven, heavy-duty, manually-operated, malediction-inducing locks to the west. The Canal & River Trust's area office stands, just across the Knights Way footbridge on the opposite bank on the site - appropriately or otherwise - of Hargreaves coal wharf.

Dominated by the Royal Armouries visitor attraction, Leeds Dock (formerly known as New/Clarence) promotes itself as a 'waterfront village and cultural hub', as evidenced by the presence of Sky's Digital & Technology Services Campus and Channel 4's weekday broadcasting from here of the talk show *Steph's Packed Lunch*. Visitor moorings with electricity (CRT pre-paid card required) and water are provided within the dock precincts, and if you're not proceeding up onto the Leeds & Liverpool Canal, it provides a practical base from which to explore the city centre, less than quarter of an hour's walk in a north-westerly direction, though the thoroughly commendable water taxi service (see page 34) lends aquatic connectivity as well.

That said, to not explore the upper reaches of the Aire & Calder, as it weaves its way beneath an interesting assortment of bridges to its assignation the Leeds & Liverpool Canal, would be something of a missed opportunity. Crown Point Bridge, the first you encounter from the south, is a Gothic confection in cast iron erected in 1842 to the design of the ubiquitous Leathers, father and son. It is followed by the Centenary Footbridge put up in 1993 to link the Brewery Wharf development with The Calls on the opposite bank. Centre-piece of the former is a rotunda which began life as an exhibition centre for Joshua Tetley's. You would have got long odds, then, for the subsequent closure and demolition of the brewery itself. Tetley's iconic Yorkshire bitter is brewed now in Wolverhampton, of all places.

Exploration pays dividends. Neighbouring streets evoke an atmosphere of Georgian Leeds, when the Aire was very much the lifeblood of trade. Dock Street leads to Victoria Quays, a 1980s redevelopment of the Aire & Calder Company's roofed-over terminal wharf where water-lilies proliferate (quite prettily) in place of keels. Similarly, ranged along the river frontage of The Calls on the river's northern side - not for nothing known as Warehouse Hill - stand an array of premises - some now converted into balconied restaurants and bars - which also resonate with the Aire's working past.

This hugely appealing reach of the Aire culminates in the single, 102ft 6ins, cast iron span of Leeds Bridge dating from 1873, though there had been a succession of earlier structures at this point since replacement of the original ferry in the 14th century. It was looking downstream from Leeds Bridge, that Atkinson Grimshaw (see also page 33 lower), painted a scene thronged with tugs and barges in 1880. The original is in the collection of Leeds Art Gallery, though irritatingly not always on display. Twin plaques on the gable end of the building adjoing the south-east corner of the bridge commemorate two historic events: that the teetotal Band of Hope was founded here in 1847; and that the first moving pictures taken with a single lens camera were made from an upstairs window by Louis le Prince in 1888. Prince vanished in odd circumstances two years later, but is generally regarded as the father of motion pictures. Next door down, fronting the corner of Dock Street, stand the former offices of the Aire & Calder Navigation, built in 1906 and retaining the company title in gold lettering with retrospective pride.

Proceeding upstream to Victoria Bridge - designed by George Leather Junior, opened in 1839, and named in homage to the new young queen - the Aire passes the Asda supermarket chain headquarters on the south bank and a sequence of modern office blocks - leavened by the 5-storey Victoria Mills of 1836 - on the north. The latter houses a ground floor brasserie belonging to Raymond Blanc. Hereabouts stood Goodall

continued overleaf: 31

River Lock, Leeds - the start of the Leeds & Liverpool Canal

continued from page 31:

Backhouse & Co's condiments works, manufacturers of six million bottles of Yorkshire Relish per annum.

Through the gracefully rusticated millstone grit arch of Victoria Bridge the unnavigable Aire - buffeting the unwary boater - enters from the right, emerging from the 'Dark Arches', a catacomb of vaults beneath the city's railway station. A weir precluded navigation upstream of this point. The Hilton Hotel (not to be confused with the Hilton DoubleTree Hotel on nearby Granary Wharf) stands adjacent to the site of the Co-op's former coal wharf where coal, delivered by barge, was unloaded and bagged for the domestic market.

Leeds & Liverpool Canal

Presided over by a sturdily handsome warehouse dating from 1776, River Lock (No.1) effects a sea change in pre-nationalisation (1948) ownership, engineering style and (not least) gauge. Boaters whose eyes are attuned to the narrow canals will consider the Leeds & Liverpool's locks huge, but if you've just come up from Castleford they'll immediately appear both diminutive and archaic. Old photographs illustrate a ramshackle conglomeration of wharves and boatyards difficult to correspond with the St Tropez ambience of the present day.

Immediately above River Lock two short arms (one with gates to act as a drydock) are all that remain of William Rider's boatyard where wooden barges were still being built into the 1950s and which finally closed in 1972. Visitor moorings (with electricity provided) are laid on at right angles to the canal at the point where an arm was constructed to connect via a subterranean lock with an upper reach of the Aire; a link principally employed by vessels serving Bean Ing Mills, the Corporation's electricity works at Whitehall, and Watson's soap works.

Overlooked by the Leeds & Liverpool Co's classically-designed Canal Office, the canal passes through Office Lock. Thankfully preserved intact, Tower Works' Italianate chimeys have nevertheless been engulfed by new apartment blocks. Milepost 0/127¼ stands alongside the towpath. Their presence is depicted on our maps counting westwards from Leeds (though admittedly without the quarter, which would have proved irksome to reproduce ad infinitum). Note the ramp cut into the canal bank, a feature of the L&L throughout its length, designed for the rescue of boat horses which inadvertently found themselves submerged in the canal.

Passing beneath the cat's cradle of railway lines, which form the station's western approaches, the canal draws alongside the River Aire across which a crescent-shaped footbridge provides useful access. The towpath here - and indeed throughout the canal corridor between Leeds and Bingley - is designated the Aire Valley Towpath Route and is deservedly popular with pedestrians and cyclists alike, being part of National Cycle Route 66.

Bridge 225G is an adjunct of Monk Bridge, a lattice girder structure spanning the Aire. The city's Coat of Arms is cast into the ironwork of the canal bridge though, unfortunately, a row of trademark owls perched atop the ironwork, has not survived into modern times. A more imposing bridge (225F) follows, designed with the chutzpah of the Railway Age to carry the lines into Central Station where Leeds folk entrained for London.

St Ann's Ings Lock lies in the early morning shadows of student apartment blocks; though by the time most of their occupants rise the chamber is likely to be basking in the full glare of the noon-day sun. Above the lock on the towpath side stands the half-mile marker out from Leeds, whilst opposite you'll catch glimpses of various railway premises, including a crescent-shaped workshop and engine roundhouse erected in the 1840s for the Leeds & Thirsk Railway. Astonishing survivals, they are well worth going up onto Wellington Road (Bridge 225D) to see at closer quarters; not least as the former is now a Majestic wine merchant outlet.

Burgeoning margins of arrowhead would not have been tolerated when this was a working waterway, but they soften the edges of a canal that no longer has commercial aspirations. When they were digging the foundations for Castleton Mill in 1836, a Civil War cannon ball was uncovered. The four-storey mill was built for flax-spinning. Today it houses numerous businesses, not least a boat hire base.

---

continued from page 29:

Sea Cadets and rowers make good use of the Aire & Calder downstream of Knostrop Fall Lock. The cylindrical, Martello like tower across from the lock, together with the stone pier rising out of the Aire, are all that remain of a gargantuan railway swing-bridge erected in the 1890s to carry the Great Northern Railway to a new goods depot at Hunslet. The line was abandoned in 1966 and the bridge was demolished eleven years later. The bridge never swung. The GNR omitted to commission any machinery.

Knostrop Cut was the subject of an ethereal oil painting by John Atkinson Grimshaw (1836-93) who lived at Knostrop Old Hall, demolished in 1960. A native of Leeds, Grimshaw went in for luminous night scenes, often of harbours, and fittingly, where this guide is concerned, painted a number of Liverpool studies. Filled with misplaced optimism, British Waterways developed warehousing alongside the Knostrop Cut in the mid-1950s. They even introduced a 'Continental Container Service' with lorry-to-barge-to-ship logistics between Yorkshire's industrial heartland and northern Europe. Hull's dockers were not amused, and the bold idea quickly languished.

Knostrop Flood Lock is merely a pair of mitred gates, left permanently open unless there is heavy flooding. Overlooked by new housing, the stubby arm represents the Aire's original course before the Knostrop Cut was dug in 1775 and the river diverted to the opposite side. Whitaker's tanker barges - some as capacious as 700 tons - discharged petroleum here until 1988. Hunslet Mills, erected in 1838 as a centre for the spinning of flax, are being redeveloped as apartments.

Allerton Bywater
Map 12

Former mining village much redeveloped with housing. Three pubs, fish & chips, sandwich shop, pharmacy and butcher. Shops along road to Great Preston. Buses to/from Leeds and Castleford.

Things to Do
BUCYRUS-ERIE - Bullerthorpe Lane. Tel: 0113 288 9088. American built 1150B opencast mining dragline open to the public on selected dates. LS26 8AL
RSPB ST AIDAN'S - Astley Lane. Visitor centre with cafe open daily from 10am. LS26 8AL

Woodlesford
Map 12

Expedient staging post between Leeds and Castleford. A pub, sandwich shop, and several takeaways in the centre of the village, seven or eight minutes walk (uphill past the former church) from the visitor moorings at Swillington Bridge.

Things to Do
LEVENTHORPE VINEYARD - Bullerthorpe Lane. Tel: 0113 288 9088. Established 1985, south-facing vines in free-draining sandy soil. Open daily 11am (noon Suns) to 4pm but best to telephone and check before setting off to walk from the A&C. LS26 8AF

Leeds
Map 14

Captain of Industry - clothing, printing, brewing & engineering - with a sentimental streak - sculpture, painting, symphony concerts and music hall - pragmatic, fortune-amassing Leeds shares more than a 19th century canal with a certain city on the other side of the Pennines. Spawned by the capital's hegemony, they share an inferiority complex, which has driven them, down the years, to vainglorious architectural statements of parity, exemplified, in Leeds' case, by Cuthbert Brodrick's Titan of a Town Hall, the same architect's elliptical Corn Exchange, the symphony in glass and iron that is the City Markets, and the dazzling Portland stone-built Civic Hall with its iconic golden owls. Any city in the world would be proud of such buildings, and Leeds can muster reinforcements at whim. Thus it is a simultaneously hedonistic and instructive place to perambulate, from its soaring Victorian and Edwardian shopping arcades - between The Headrow and Briggate - to oases of quietude such as Park Square, a backwater in what was originally a district of discerning domesticity but which is now populated by doctors, dentists and lawyers. The leafy square is dominated by St Paul's House, the Moorish fantasy of a fortune-amassing pioneer in ready-to-wear clothing. In short, Leeds has nothing to feel inadequate about.

Eating & Drinking
AMBIENTE - The Calls. Tel: 0113 2461848. Tapas bar specialising in Spanish seafood. LS2 7EW
BRASSERIE BLANC - Sovereign Street. Tel: 0113 220 6060. Reliable Raymond Blanc restaurant in former riverside linseed mill dating from 1836. LS1 4BJ
FRIENDS OF HAM - New Station Street. Tel: 0113 242 0275. Mouth-watering cured meat and cheese platters accompanied by locally procured beers and even Leventhorpe (Map 12) wine. LS1 5DL
MUMTAZ - Leeds Dock. Tel: 0113 242 4211. Dockside Indian restaurant from 4pm ex Mondays. LS10 1PJ
WHITELOCKS - Turks Head Yard (off Briggate). Tel: 0113 242 3368. On the CAMRA national inventory of historic pub interiors. Atmosphere, ales and inexpensive nourishing meals. Outdoor patio. LS1 6HB

Shopping
In the absence of city-centre supermarkets, food shoppers have to think like their forebears and revert to independent retailers. Nowhere better to start than Kirkgate Markets whose fishmongers are exceptional.

Things to Do
TOURIST INFORMATION - The Headrow (Art Gallery). Tel: 0113 378 6977. LS1 3AA
ARMLEY MILLS - canalside bridge 225 (Map 15). Tel: 0113 378 3173. Open daily (ex Mon). Displays of industrial history in former woollen mill. LS12 2QF
ART GALLERY - The Headrow. Tel: 0113 378 5350. Admission free. Fabulous Tiled Hall cafe. LS1 3AA
CITY MUSEUM - Millennium Square. Tel: 0113 378 5001. Admission free. LS2 8BH
LEEDS MINSTER - Kirkgate. Tel: 0113 245 2036. St Peter's is a substantial Victorian church near Crown Point Bridge. Cathedral status eluded it, the Diocese honouring Wakefield instead. Yet the effect is sumptuous, particularly the interior which contains a 10th century stone cross. Refreshments. LS2 7DJ
ROYAL ARMOURIES - Armouries Drive (adjacent Leeds Lock). Tel: 0113 220 1916. Admission free. Military history, jousting, falconry etc. LS10 1LT
THE TETLEY - Hunslet Road. Tel: 0113 320 2323. Art gallery, bar & kitchen housed in Art Deco offices of former brewery to south of Leeds Bridge. Open 10am-5pm ex Mon & Tue. LS10 1JQ
THWAITES WATERMILL - canalside below Knostrop Fall Lock (Map 13). Tel: 0113 378 2983. Water powered mill and blacksmith's workshop. Open Sat and Sun 12-4pm. LS10 1RP

Connections
TRAINS - direct access to the station from Granary Wharf via south entrance. Tel: 0345 748 4950.
BUSES - service 29 operates at 20 minute intervals from stop A1 at Leeds Dock into the city centre and bus station.
TAXIS - City Cabs. Tel: 0113 246 9999.
WATERBUS - operates 10am-6pm daily between Leeds Dock and Granary Wharf. £1 per ride.

GALLERY

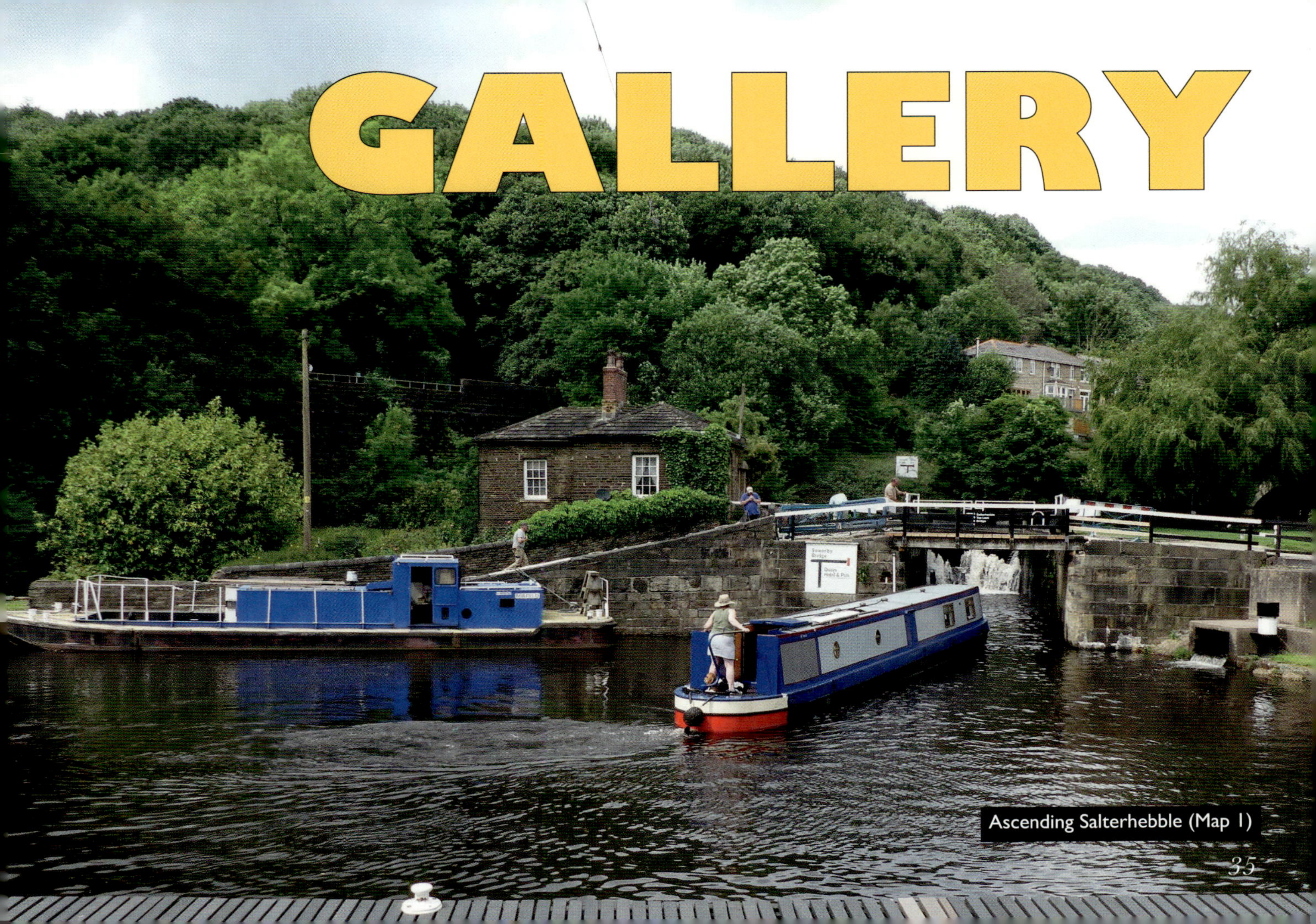

Ascending Salterhebble (Map 1)

'Nearly There!', Brighouse (Map 2)

ROCKABILLY
G17210

'Unwillingly to School', Shipley (Map 17)

38

John's Drum, Bingley 5-Rise (Map 17)

Springs Branch Trip Boat, Skipton (Map 20)
LEO No 1
PENNINE CRUISERS
BOAT TRIPS
30min BOAT TRIP ON N B LEO
BEHIND & UNDER THE CASTLE
ALL SEATS £4.00
LEO No 1

Retired Short Boat at Greenberfield (Map 23)

'Working from Home'
Slater Terrace, Burnley (Map 25)

A Moment for Reflection, Withnell Fold (Map 29)

Sprint Finish! Johnson's Hillock Locks (Map 30)

Damp Day at Wigan Pier (Map 32)

Dream Catcher locking down the Rufford Branch (Map 34)

Sonia Rose

LIKE an office on a Dress-Down-Friday, the Leeds & Liverpool appears misleadingly informal as it threads its way through the Aire Valley, barely ruffled by the urbanisation which surrounds it. Hooliganism, both real and rumoured, is an unsavoury fact of life. As a consequence, the Canal & River Trust limit passage between Kirkstall Lock (No.7) and the top of Newlay staircase (No.13) to the hours of 8am to 4pm, March to October*, during which the locks involved are overseen by CRT staff and volunteers. All things considered, it's a wise enough precaution, though it does rather rush you through what ought otherwise to be a rewardingly entertaining length of canal.

Completed in 1777 - fortuitously just before the company temporarily ran out of capital - this section of the canal features three sets of riser or staircase locks, a method of construction which, in hindsight, was shown to be both time-consuming and wasteful of water. Characteristics which pertain to this day. Nevertheless, they add variety to the modern-day boating experience and you will soon get the knack of operating them - moreover, volunteer lock-keepers are often on hand to help you through the three rises. Overlooked by Maurice Dixon's (erstwhile manufacturers of flannels & tweeds, serges & suitings) watertower, and enlivened by faded murals, Oddy 2 Rise sets the tone, the keeper's former accommodation balefully boarded up. Inscrutable distribution premises have replaced the works of some of the city's pre-eminent industrial enterprises: Greenwood & Batley's Albion Works churned out everything from machine tools to First World War tanks; at Leeds Forge they made ship boilers and rolling stock for the world's railways; and in Leedham & Heaton's Iron Works they specialised in spades and shovels.

Spring Garden Lock precedes milepost 1 as the canal runs along a sandstone shelf above the Aire: six locks up and you are already considerably

*8am-3pm in winter by prior arrangement with CRT.

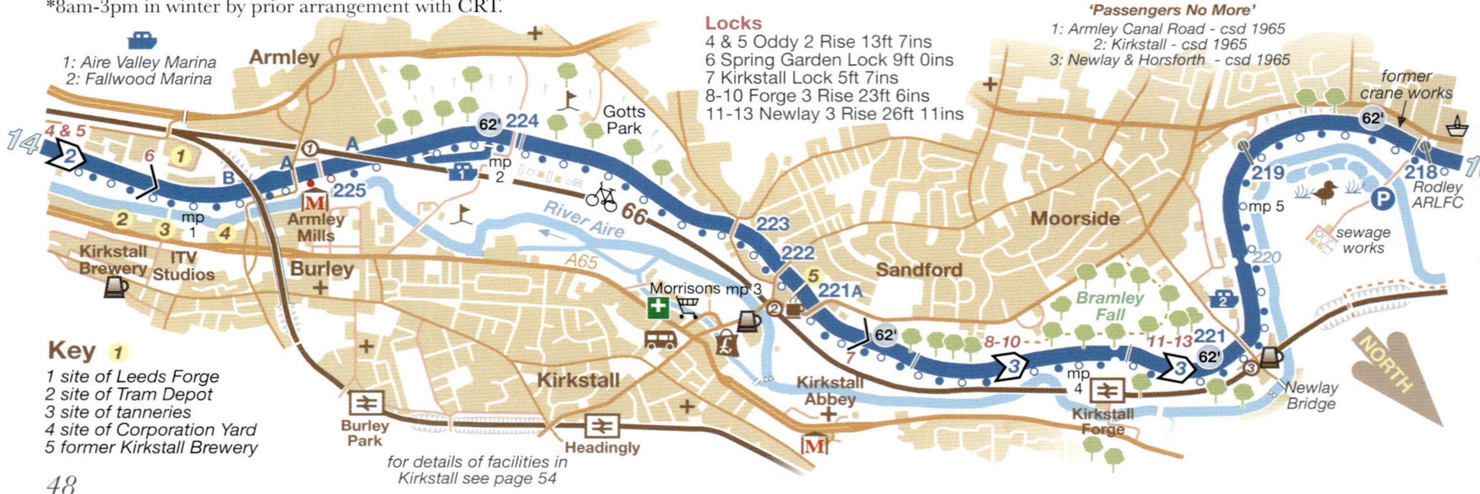

higher than the river. Bridge 225B is part of an impressive railway viaduct which carries the Harrogate line across the valley. Note how the railway builders were determined to stamp their superiority over the canal in no uncertain architectural terms, as if to say: 'Watch it mate, we're the future, you're obsolete.'

Situated to harness the power of the Aire, Armley Mills predated the canal but must have been given a fillip when it arrived on the scene. The buildings that exist today date largely from the beginning of the 19th century following a serious fire, an occupational hazard where milling is concerned. Benjamin Gott ('no one in the West Riding stood higher as a man of business' waxed his obituary in 1840) instigated a revolutionary form of fireproof reconstruction and installed gas lighting and the mill was successfully employed in the manufacture of woollens up until 1969 whereafter it was purchased by Leeds City Council and eventually re-opened as a museum; and a very entertaining one at that.

Hefty 'Pratt Trusses' carry the once four-track Midland Railway's line to Skipton and beyond over the canal before you come upon an intriguing side dock used now for private moorings. This was the unloading lay-by for Kirkstall Power Station, busy with discharging short boats up until the 1960s. As at Skelton Grange (Map 13), the bollards provide eloquent reminders of forgotten trade. Rail-served in later years, the generating plant closed in 1978. On the opposite bank, Gotts Park is a golf course now, the eponymous Benjamin's mansion providing a palatial 19th hole. Gott acquired the property in 1803 and called in Humphry Repton for advice on improvements to the house and grounds which became his principal residence; a residence, moreover, whose interiors were adorned with paintings by the likes of Titian, Rubens and Caravaggio. Repton's subtle landscaping afforded vistas of Armley Mills and Kirkstall Abbey.

The original Kirkstall Brewery has been converted to provide accommodation for Leeds Metropolitan University students, but at one time the canal played a significant role in transporting the brewery's finished product. Hogsheads of beer were conveyed by barge to Goole and transhipped for export aboard the brewery's own steamships *Charente* and *Kirkstall* which traded as far afield as the Antipodes. The brewery closed in 1983, but a new concern began trading under that name in 2011 and, after a period of brewing nearby, moved to larger premises on Kirkstall Road, near where ITV's Emmerdale studios are (see map).

18th century Airedale was a virtual Arcady. Artists no less illustrious than Joseph Mallord William Turner were attracted by the romantic decay of Kirkstall Abbey - as also were John Sell Cotman and Thomas Girtin. Turner painted Kirkstall Lock in 1824, and it's too tempting not to express the flippant opinion that it hasn't been painted since. The abbey's ruined stonework - still smoke-blackened after years of exposure to the valley's heavy industry - can be glimpsed beyond the railway; though some local trick of the light seems to rarely thrust it forward.

And so to those pesky 3-rises, reputedly the focus of perverted entertainment for the youth of neighbouring housing estates. Acts of hooliganism tend to expand like heated metal with repeated telling, but received wisdom suggests that the earlier in the day you negotiate these locks, the less angst you'll experience. Between the locks the canal skirts the rocky terrain of Bramley Fall, formerly an area of extensive quarrying. Bramley Fall gritstone was employed in the building of Kirkstall Abbey, but quarrying was at its zenith in the 19th century when the stone's proven durability and resilience to water saw its use in construction - amongst much else - of the Aire & Calder Navigation, London Bridge, and Martello Towers along the east and south coasts to counter the threat of a Napoleonic invasion. The quarries were worked out early in the 20th century and are now in use as public woodland. Opposite, though largely masked by woodland, the site of Kirkstall Forge - abandoned in 2003 after four centuries of continual production, latterly motor vehicle axles - is being redeveloped for mixed use, complete with a new railway station.

On no account eschew a short detour from Bridge 221 to John Pollard's exquisite iron bridge, cast at Shelf Iron Works at Shelf (see also Map 1) in 1819. Back on the canal, Bridge 219 is the first of the Leeds & Liverpool's many swing-bridges encountered when journeying north and west. Familiarity will breed not so much contempt, as a deep and lasting loathing.

SOME gentle Delius might provide a delightful soundtrack to your exploration of these waters. The Bradford born (1862-1934) composer's limpid melodic lines would form a fitting accompaniment to the canal's ethereal progress. Waterside dwellings of Rodley and Apperley Bridge notwithstanding, the *mise en scene* of this section of the Leeds & Liverpool is predominantly rural, and when we state that the highlight of this section is the vast sewage works at Esholt, it is for you to decide how firmly Pearsons have their tongue in their cheek.

Six miles out from Leeds - though six hours cruising on account of thirteen locks - Rodley is regarded as a useful pitstop in most boating itineraries. Certainly there are two good sets of visitor moorings, though you will need to be well victualled, for there is no food shop in the vicinity. Rodley was always known for its crane and excavator works. Plant with Smiths of Rodley proudly embossed on its metalwork has performed heroics at some of the world's greatest civil engineering projects, from the Manchester Ship Canal to the Aswan Dam.

Passing beneath the concrete span of the A6120 - aka Leeds Ring Road - adds conviction to the feeling that you have finally escaped the city's clutches. Just as we exhorted you to go and see John Pollard's bridge at Newlay, so a short stroll from swing-bridge 216 will lead to a packhorse bridge over the Aire: treasures that non-Pearson users pass ignorantly by.

Between mileposts 7 and 8 the railway drops by for a chat, its diet of frequent electric trains to Bradford, Skipton and Ilkley, leavened with diesel units bound for the famous Settle & Carlisle, and goods trains, rumbling south with Ribblehead granite or Grassington limestone. Would that the canal had some trade to boast about, other than of the leisure kind.

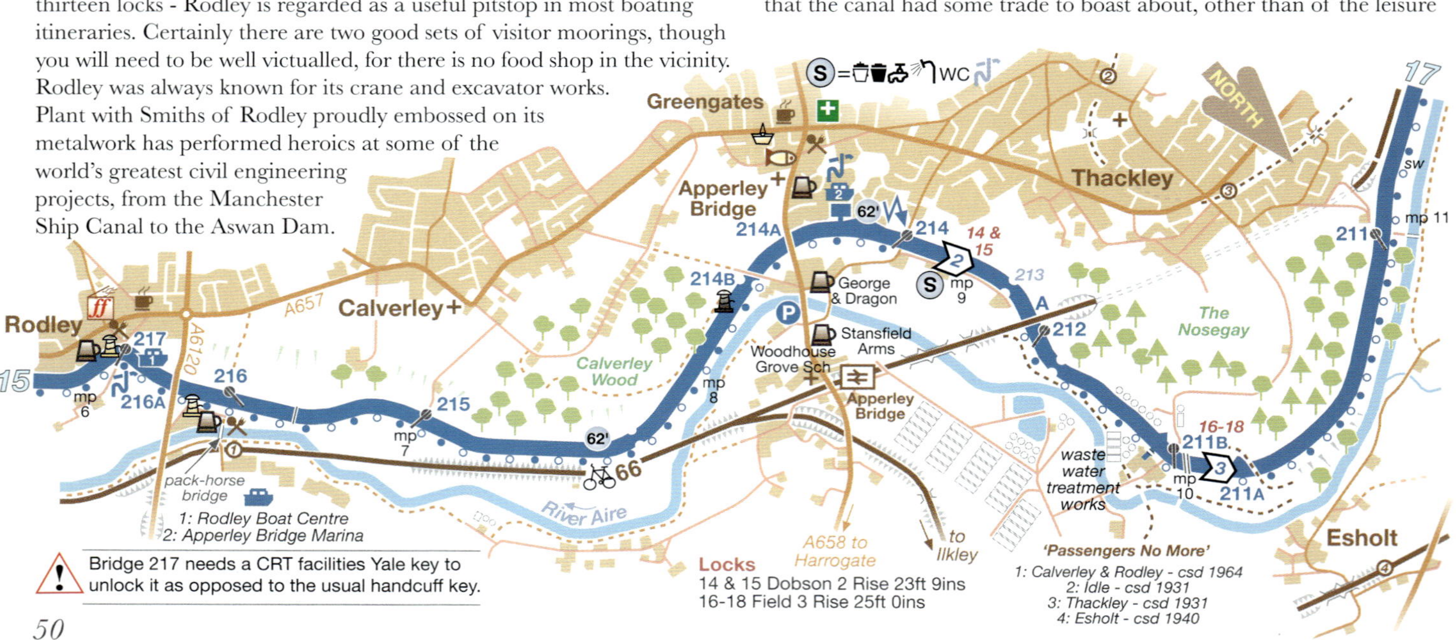

variety. Cargo carrying on this side of the Pennines had more or less died out by the end of the 1950s, Skipton gas works being one of the last recipients of regular coal deliveries from Park Hill Colliery on the outskirts of Wakefield (Map 6). In the early Eighties effluent was carried from Esholt to Knostrop, but that operation was relatively short lived. By the way, the railway milepost reading 202 records the distance from London St Pancras, for this, of course, was once the mighty Midland Railway.

Leaving the railway to follow a more direct course - bridging the river twice before tunnelling under high ground known as The Nosegay - the canal skirts Calverley Wood where old quarry faces once provided work for indigenous short boats. Woodhouse Grove School's playing fields run down to the banks of the Aire. Here, one wintry afternoon in 1969, Lock-wheeler scored a try from a five yard tap penalty; simultaneously acquiring the sobriquet 'Steel-head'. Bradford City FC train alongside the river.

Apperley Bridge Marina is based on a housing development co-funded by British Waterways before they became the Canal & River Trust. The stone built houses are pleasing to the eye and harmonise well with the adjoining textile mill. Schemes like this may not appear at first to be core business activities, but they help to top up the coffers, and have become increasingly important now that the canal system is administered by a charity. Alongside Dobson's 2 Rise locks stand CRT's maintenance yard along with a row of cottages built by the L&L to house their employees. Strange to think that all these miles of canal are cared for from this modest collection of workshops and cottages.

And so to that promised highlight - Esholt Waste Water Treatment Works. Irritated by continual complaints from the owners of Esholt Hall that the Aire running through their grounds was polluted beyond forebearance, Bradford Corporation compulsorily purchased the estate and set about turning it into the largest sewage works in Europe. It opened in 1924, complete with its own canal basin, beside which Grecian inspired 'press houses' bore the city's coat of arms and motto LABOR OMNIA VINCIT - something which Lock-wheeler's Latin master was forever impressing upon him; to little avail. The site - operated by Yorkshire Water nowadays, who pointedly avoid use of the term 'sewage' in its title - once boasted 22 miles of internal railway shunted by little saddle tanks (unambiguously nicknamed 'Pongos') ecologically powered in advance of their time by a derivative of human waste.

Use of Field 3 Rise Locks (Nos. 16-18) is restricted to the hours of 8am to 4pm, though they aren't necessarily manned. Milepost 10 (yes, we know, and a quarter!) was reinstated as part of the laudable 'Every Mile Counts' initiative marking the bi-centenary of the Leeds & Liverpool Canal's completion in 1816.

Rodley Map 16

Popular spot for a breather before or after Leeds.

Eating & Drinking

CAFE FRAICHE - Rodley Lane. Tel: 0113 236 3520. Well-appointed little cafe open 9am-3pm daily. Eat in or take-away paninis etc. LS13 1LB

EPHESUS - Rodley Lane. Tel: 0113 256 1668. Turkish restaurant open daily from noon. LS13 1HU

THE RAILWAY - adjacent Bridge 216. Tel: 0113 819 7181. Cosy pub between railway and canal. Food served from noon throughout daily. LS13 1PY

RODLEY BARGE - Town Street (adjacent Bridge 217).

Tel: 0113 257 4606. Stone-built pub overlooking the canal. Lunches from noon daily. LS13 1HP
Also another pub (The Owl) and Balti take-away.

Connections

BUSES - services 670/1 operate (ex Suns) half-hourly to/from Leeds and Bradford. Tel: 0871 200 2233.

Apperley Bridge Map 16

Former mill village on the Bradford to Harrogate road.

Eating & Drinking

ALDO'S - Harrogate Road. Tel: 01274 270802. Italian restaurant closed Mon & Tue. BD10 0RB

STANSFIELD ARMS - Apperley Lane. Tel: 0113 250 2659. Large 16th century pub serving bar/restaurant food 5 mins walk to north of canal. Food served from 10am daily throughout. BD10 0NP
The George & Dragon is part of Vintage Inns.

Shopping

Small Asda and pharmacy adjacent crossroads about 7/8 minutes walk from Bridge 214A.

Connections

BUSES - service 747 operates hourly to/from Bradford and Leeds/Bradford Airport. Tel: 0871 200 2233.

TRAINS - half-hourly trains to/from Leeds etc.

for details of facilities in Shipley and Saltaire see page 54

for details of facilities in Bingley see page 56

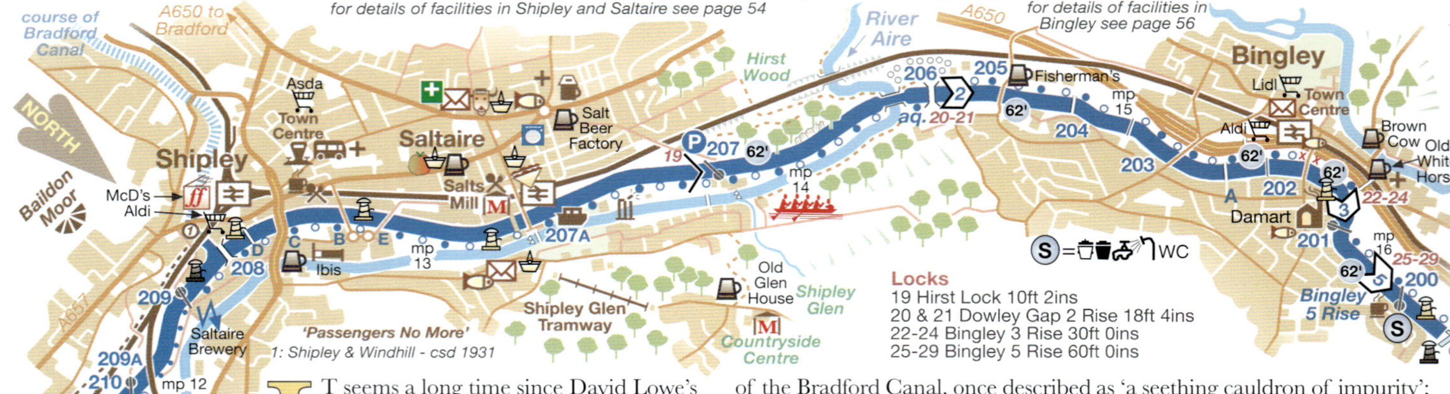

IT seems a long time since David Lowe's Metro Waterbus plied these waters. We miss the 'bus stops' which stood so incongruously canalside, as though any minute a blue & cream Bradford Corporation half-cab might come wheezing along the towpath, advertising Crawford's Cream Crackers or Vernons Pools, whilst absorbing and disgorging passengers en route. A trip boat offers 30 minute cruises from Saltaire, but that's hardly the same as a timetabled service from Shipley to Bingley. Why do worthwhile initiatives always seem to slip through our collective fingers, like jelly escaping from a warm pork pie?

Twelve miles out from Leeds you bump into Shipley, an unprepossessing town on the surface of it, which you may warm to on further acquaintance: the canalside presence of Saltaire Brewery providing a reassuring outlier to the town. Junction Mills - by Bridge 208 - recall the former egress of the Bradford Canal, once described as 'a seething cauldron of impurity'; much like the House of Commons, then. It was reputedly so polluted by waste from mills and dyeworks that the surface flickered with a blue gas flame. Limited 24 hour visitor moorings are signposted by Bridge 207D, but the best moorings (of a generous 7 days duration) are at Ashley Lane, on the offside between bridges 207B and 207E. Indeed this is the best bet if you are planning to stay overnight in the vicinity of Shipley or Saltaire, for the latter's designated site (south of Bridge 207A) is limited to daytime only, perhaps in deference to the Bradford District NHS Care Trust's occupancy of the former mill building alongside.

Shipley's massive canal warehouses create a distinct impression by virtue of their size alone. Once they would have been stuffed to the gills with angora fleeces. Their simple dignity contrasts with futuristic office blocks strung along the opposite bank of the canal: architecture oscillating between the centuries and getting a black eye in the process. Shipley's old Midland Railway station boasts an unusual triangular layout. Stroll up its cobbled approach road and treat yourself to a short, but enjoyable excursion by electric train to Ilkley or Bradford, both worthwhile places to visit.

Manual swing-bridges will need a CRT handcuff/T key to unlock them. Electrified Bridge 209 requires a CRT facilities Yale key to access its control panel.

Once you have grown accustomed to Saltaire, there is little imperative to go anywhere else - in the world. Bournville, Port Sunlight and New Lanark might beg to differ, but it seems to us that this is the most cohesive industrial model village in the land, the most perfectly realised workers' paradise. It deserves to be savoured, as perhaps Sir Titus always intended: its canyon of fudge-coloured mills, elegant United Reform church, and streets of dignified worker's housing named after Titus Salt's children and children's children. As a blueprint for post-industrial regeneration, Saltaire appears unsurpassed.

Seemingly reluctant to leave, the canal proceeds past Salts' sports grounds (tennis, bowls, cricket, football) towards the glacial moraines of the Dowley Gap, limbering up for the Bingley Five Rise with a hop, skip and a jump in the form of a single, double and triple lock. A woodland interlude and an aqueduct add to your momentum. The latter spans the Aire along a reach where otters have been reintroduced, reputedly colonising the canal. Spraints were not immediately evident when Lock-wheeler - rowing bow in an eight - went through his watery paces at the Bradford Regatta of 1969.

Shoulder-barged by the 'award-winning' (sic) A650, known hereabouts as Sir Fred Hoyle Way - in honour of the astronomer, born near Bingley in 1915 - the canal takes a deep breath and squeezes through what little space is left. The civilised world is divided into those who know Bingley as a centre of thermal underwear manufacture, and those who recognise it as the location of one of Robert (founder of the IWA) Aickman's 'Seven Wonders of the Waterways', the gargantuan Bingley Five Rise staircase locks. Boaters arriving from the south have the benefit of a warm-up act in the shape of a self-effacing 3 Rise, and must make themselves known to the amenable band of lock-keepers who reside in a bothy at the top of the 5 Rise. Both staircases operate to a timetable (well, sort of), so be prepared to exercise patience. Passage through them is restricted to the hours of 8am to 4pm (3pm out of season). The 3 Rise, loomed over by Damart's mill, and offering a view across the dual-carriageway of an old Midland Railway warehouse, is entertaining enough, but it is

Instruction notices at Dowley Gap 2 Rise quaintly refer to 'penning', whilst the keeper's house fondly recalls the era when they regularly won best-kept-length competitions. Who knows now what a BKL even looks like? Bridge 204 carries the pipework of Bradford Corporation's Nidd Aqueduct water supply channel. The trio of twelve-storey tower blocks which hitherto heralded Bingley, have been demolished. They only lasted fifty years. Good riddance! Less intrusive housing is to be built on the site. the 5 Rise which draws the crowds. Altering the level of the canal by some sixty vertiginous feet, it is a cathartic experience which the canal takes more or less in its stride, even if it leaves most canal users reeling. The staircase has stood its ground since 1774 and must be credited to John Longbotham, the L&L's chief engineer. In spring the aroma of wild garlic permeates your passage through the 5 Rise, its fragrant source can be found skulking behind a shelter beside the by-wash of the bottom chamber.

Kirkstall
Map 15

Leafy suburb of Leeds, cheek by jowl with Headingley and its immortal cricketing associations: - 'and Trueman comes in to bowl from the Kirkstall Lane End' - and both codes of rugby too. The prominent spire of St Stephen's is the work of Richard Chantrell (1793-1872) who also designed Leeds Minster. Royalists and Parliamentarians fought over the river bridge during the Civil War. Scenes from Alan Plater's TV series *The Beiderbecke Affair* were shot locally.

Eating & Drinking
BRIDGE INN - Bridge Road. Tel: 0113 278 4044. Well-appointed riverside pub open from noon daily. Good choice of food served throughout. LS5 3BW
HOLLYBUSH - community focussed cafe, shop and garden by Bridge 221A, entrance off towpath.

Shopping
Morrisons supermarket, Boots pharmacy and retail park (The Range, Holland & Barrett, Iceland etc).

Things to Do
ABBEY HOUSE MUSEUM & KIRKSTALL ABBEY - Abbey Road, Kirkstall. Tel: 0113 378 4079. Closed Mondays. Victoriana and monastic ruins. LS5 3EH

Shipley
Map 17

It would not be politically correct these prickly days to refer to Shipley as Saltaire's ugly sister. In any case, to do so would not be strictly accurate - not professional guidebook compilation - for there is an austere retro beauty to be admired about Shipley, especially in the vicinity of hilltop Market Square, an example of Sixties 'brutality' which architectural students probably come from far and wide to see.

Eating & Drinking
INTERLUDE - Westgate. Tel: 01274 809636. Charming tea room open 11am-4pm daily ex Sun. BD18 3QX

WATERSIDE - Wharf Street. Tel: 01274 594444. Bistro & bar in old canal warehouse. Open Tue-Sat 10.30am-3pm, 5pm-9pm. BD17 7DW

Shopping
Aldi lurks handily beneath the canal embankment by Gallows Bridge (207D), but you should really make a point of visiting the Market Square and Market Hall. The open market operates on Mon/Fri/Sat, the semi-subterranean indoor hall daily ex Sun. There is an Asda on the far side of town.

Things to Do
SALTAIRE BREWERY - Dockfield Road. Tel: 01274 594959. Canalside brewery established in 2005. Shop offering casks, bottles and gifts open daily. Brewery Tap bar open from noon Thur-Sun. BD17 7AR

Connections
TRAINS - frequent services to/from Leeds, Bradford, Skipton (via Saltaire) and Ilkley. Tel: 0345 478 4950.
TAXIS - AA. Tel: 01274 584444.

Saltaire
Map 17

Many will consider Saltaire's World Heritage Site status an unnecessary accolade, for this heavenly model village is quite capable of existing on its own merits, remaining recognisably as its creator, mid nineteenth century mill owner and philanthropist, Sir Titus Salt, intended it to be. Post-industrial Saltaire was reinvented by another visionary, Jonathan Silver - sadly himself deceased - and Salts Mill is just as much his monument as Titus Salt's, illustrating to perfection what might be done with all those crumbling dinosaurs in Burnley, Blackburn, Wigan et al if similarly inspired imagination and entrepreneurial skills were to be employed. In motor-biking circles, Saltaire is revered as the home, up until 1950, of Scotts Motorcycles. In trolleybus circles, it was part of the Bradford network, the last to lower its poles in 1972. Look out for the four lions outside Victoria Hall - anything Trafalgar Square can do ...

Eating & Drinking
OLD GLEN HOUSE - Prod Lane, Baildon. Tel: 01274 597777. Country inn accessible via Shipley Glen Tramway or local public footpaths. Food served Wed & Thur 4-8pm, Fri 12-3pm & 5-8pm, Sat 12-8pm and Sun 12-6pm. BD17 5BN
SALT BEER FACTORY - Bingley Road. Tel: 01274 582111. Ossett micro-brewery in old tram depot. Open from noon, food served throughout. BD18 4DH
SALTS DINER - Victoria Road. Tel: 01274 530533. Cafe/restaurant on first floor of gigantic canalside mill; wonderful food, wonderful setting. Open Wed-Sun from 10am, last orders 4.30pm. BD18 3LA

Shopping
You'll find difficulty in dragging some of the crew away from Salts Mill (open Wed-Sun) with its array of retail outlets; including one of the best bookshops we know. Yorkshire Curd Tarts (amongst other myriad delights) from Salts Village Bakery just up from Bridge 207A. Launderette on Saltaire Road (A657). A little bit further on, Bingley Road boasts an array of shops, including a Co-op, bakery, pharmacy, post office and Binns award-winning butchery/deli.

Things to Do
1853 GALLERY - Ground Floor, Salts Mill. Tel: 01274 531163. Open Wed-Sun. Exhibits include premier collection of local artist made good, David Hockney. Visit also 'Gallery 2' and the History of Saltaire exhibition. BD18 3LA
SHIPLEY GLEN TRAMWAY - 5 minutes stroll across the Aire from Bridge 207A. Tel: 01274 589010. Enchanting Victorian cable tramway. Ring to ascertain post-pandemic operating times. BD17 5BN

HAPPY-GO-LUCKILY lock-less (for seventeen lazy miles) the canal slots snugly into a ledge above the Aire's suddenly constricted valley, blind to all the blandishments that the town of Keighley can offer to become better acquainted. To make the main line go through Keighley would have necessitated two bridgings of the Aire, an unnecessary expense, so warehousing was provided at Stockbridge. Nevertheless a branch to Keighley was promoted in 1821, but it never materialised. What a shame! It would have been nice to cruise across the river and thread one's way past the back alleys, mills and dyeworks of the town to a basin alongside the Keighley & Worth Valley Railway: past meeting present meeting past again, so to speak. Keighley was the home of Widdop & Co, who, at the Invincible Engine Works on Greengate, manufactured boat engines which were, to the north, what Bolinders were to the midland canals, and characteristically emitted an onomatopoeic exhaust sound: widdop, widdop, widdop ...

Riddlesden's rooftops provide counterpoint to moorland ridges beyond Keighley, and the cleft that marks the Worth Valley: Sowerby Bridge, four or five days back by boat, lies a mere dozen miles across the hills. Riddlesden's waterside gardens evoke a replete domesticity. Streams - or, rather, becks up here, we'll be bound - burrow beneath the canal on their way to the Aire, like excited children barrelling down a hillside.

Those warehouses still stand canalside either side of Bridge 197: stone to the east, brick to the west; and stuffed with people now, as opposed to wool. The newer, brick warehouse dates from the 1930s when the canal could still just about compete with the railway to win the hearts and minds of the West Riding's transport managers. Look closely at the corner of the stone-built warehouse, and you'll see a semaphore signal, once used to convey messages to the passing boatmen of non-stop fly boats during the heyday of traffic on the canal. Stockbridge to Foulridge (Map 24) was considered a day's work for a boat horse. In leisure-boating terms, it's well nigh impossible to come to terms with such feats.

Wool and sugar were still being carried over the summit from Liverpool docks to these extensive premises in the early Fifties, an activity rudely *continued overleaf:*

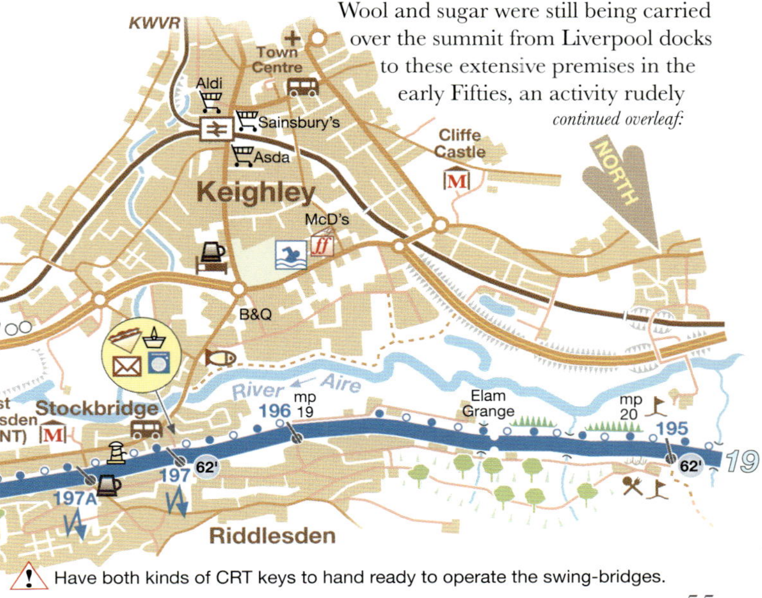

⚠ Have both kinds of CRT keys to hand ready to operate the swing-bridges.

continued from page 55:

interrupted on 17th May 1952 by a massive breach in the canal's bank which all but washed away the local golf course. Luckily, three maintenance men, already investigating a reported leak, had enough presence of mind to leap clear of the biblical deluge which ensued. Even more fortunately, they were able to insert stop planks into narrows built during the Second World War in case of bomb damage, so confining the effect of the breach to one instead of seventeen miles. Spectators poured in from all over the West Riding almost as rapidly as the water had poured out of the canal. Ten 'expert puddlers' were sent over from Wigan to reline the canal bed. They must have walked straight out of a Monty Python sketch. Ten clones in hobnail boots and braces elbowing their way like gladiators through the hushed throng - 'It's t'puddlers!' - and proceeding to go about their routine like a Morris dance in slow motion amidst frenzied outbursts of applause from a crowd hardly able to contain themselves.

Bingley   Map 17

It seems a long time since the Bradford & Bingley Building Society and their bowler-hatted icons were a household name, and thankfully their carbuncle of an office block, which so blighted the High Street, has vanished too. Of more pleasing aspect is the riverside park, an interesting and substantial parish church, and an old Butter Cross thought to date from 1212 when King John first granted a market charter.

Eating & Drinking

BROWN COW - Ireland Bridge. Tel: 01274 564345. Comfortable Timothy Taylor dining pub refurbished after floods and reached via 13th century bridge. Food served lunchtime & evenings (from 5pm) Mon-Thur, and from noon throughout Fri-Sun. BD16 2QX
OLD WHITE HORSE - Old Main Street. Tel:01274 921930. Ancient white-washed inn between the river and the church. Food daily throughout. BD16 2RH
THE FISHERMAN'S - Dowley Gap (Bridge 205). Tel: 01274 510778. Canalside pub. Open from noon daily. Food served Wed-Sun throughout. Keighley brewed Timothy Taylor on tap. BD16 1TS
STATION MASTERS HOUSE - Park Road. Tel: 01274 081850. Seafood bar/restaurant/take-away easily accessed from off-side of Bridge 202. BD16 4JD

Shopping

Aldi & Lidl. Market on Wed & Fri. Damart factory shop in Bowling Green Mills, 10am-4pm ex Sun.

Connections

TRAINS - frequent local services linking Bingley with Skipton, Saltaire, Shipley & Leeds. Tel: 0345 748 4950.

Stockbridge   Map 18

Stockbridge provides typical suburban facilities: a Co-op food store (with cash machine), post office, and a handy launderette. The Stockbridge Arms is a dine-in/take-out fish & chip shop just acrosss the Aire. Open daily (ex Mon) from 11.30am - Tel: 01535 444455.

Things to Do

EAST RIDDLESDEN HALL - Bradford Road, Riddlesden. Tel: 01535 607075. 17th century manor house, originally the home of a cloth merchant, and tithe barn surrounded by fishponds. Lovely furniture and embroideries. Operated by the National Trust. Gardens feature a herb interpretation centre. Open from 10.30am to 4.30pm mid February to end October (ex Thur/Fri). Shop and tea room. BD20 5EL

Connections

BUSES - Keighley & District 'Shuttle' 662 (amongst others) provides a 15 minute (20 mins Suns) frequency service to/from Keighley. Tel: 0871 200 2233.

Keighley   Map 18

Unexpectedly substantial town lurking a mile across the Aire from the canal. Revered in beer drinking circles as the home of Timothy Taylor prize-winning ales, and amongst railway enthusiasts as the junction for the Worth Valley Railway which will carry you nostalgically up into Bronteland and the timeless haunts of "The Railway Children". Tel: 01535 645214. Also well worth a visit is Cliffe Castle Museum housed in a Victorian textile baron's mansion on a hill to the north of the town. Tel: 01535 618231 - BD20 6LH
Swimming at the Leisure Centre on Hard Ings Road. Tel: 01535 618585 - BD21 3JN

Silsden   Map 19

An amiable little spot attractively watered by a tributary of the Aire. They're still talking about the day the 2014 Tour de France passed through.

Eating & Drinking

CURRY CORNER - Kirkgate (adjacent Bridge 191A). Tel: 01535 656222. Deservedly popular Indian. Open daily from 4.30pm, eat in or take-away. BD20 0AL
STEFANO'S - Kirkgate (top end). Tel: 01535 658555. Small but genuinely atmospheric Italian. BD20 0PB

Shopping

Friendly independents: post office, baker, butcher etc.

Connections

BUSES - Keighley & District service 62 operates half-hourly (hourly Sun) to/from Keighley and Ilkley (nice excursion 'ashore'). Tel: 0871 200 2233.
TRAINS - 15 mins frequency Airedale Line services (30 mins Suns) from Steeton & Silsden station 1 mile south of Bridge 191A. Tel: 0345 748 4950.

ONCE entirely of the West Riding, but forced since 1974 to negotiate a contrived boundary between West and North Yorkshire, the Leeds & Liverpool Canal finds Airedale at its most amiable, evoking visions of curly-haired terriers preened to perfection at Crufts. Every well-groomed dog has its day. Likewise this gorgeous canal, which conjures up one exhilarating view after another: Brunthwaite Crag to the north; Steeton Moor to the south. Down south, towpaths are chaperoned by hedges. Up north, there's no nonsense, with a drystone wall to curb youth's amatory instincts. A prominent landmark to the south is the Jubilee Tower, erected to mark Queen Victoria's 60th anniversary on the throne by the owner of Cliffe Castle (Map 18). Half folly, half home to his gamekeeper, it remains inhabited to this day.

Silsden is a handy point for re-stocking the galley from friendly little throw-back shops where you're still expected to reveal your life history over this grocery transaction or that. Time was when Silsden Co-op's barge, *Progress*, worked hard on the canal fetching and carrying produce vital to the little mill town's economy. Nowadays the main street reverberates to juggernauts and the canal wharf has become the point of departure for boating holidaymakers. Call that 'progress' if you like, but it is good to see Silsden Boats offering wide-beam boats for hire on what is after all a wide-beam waterway.

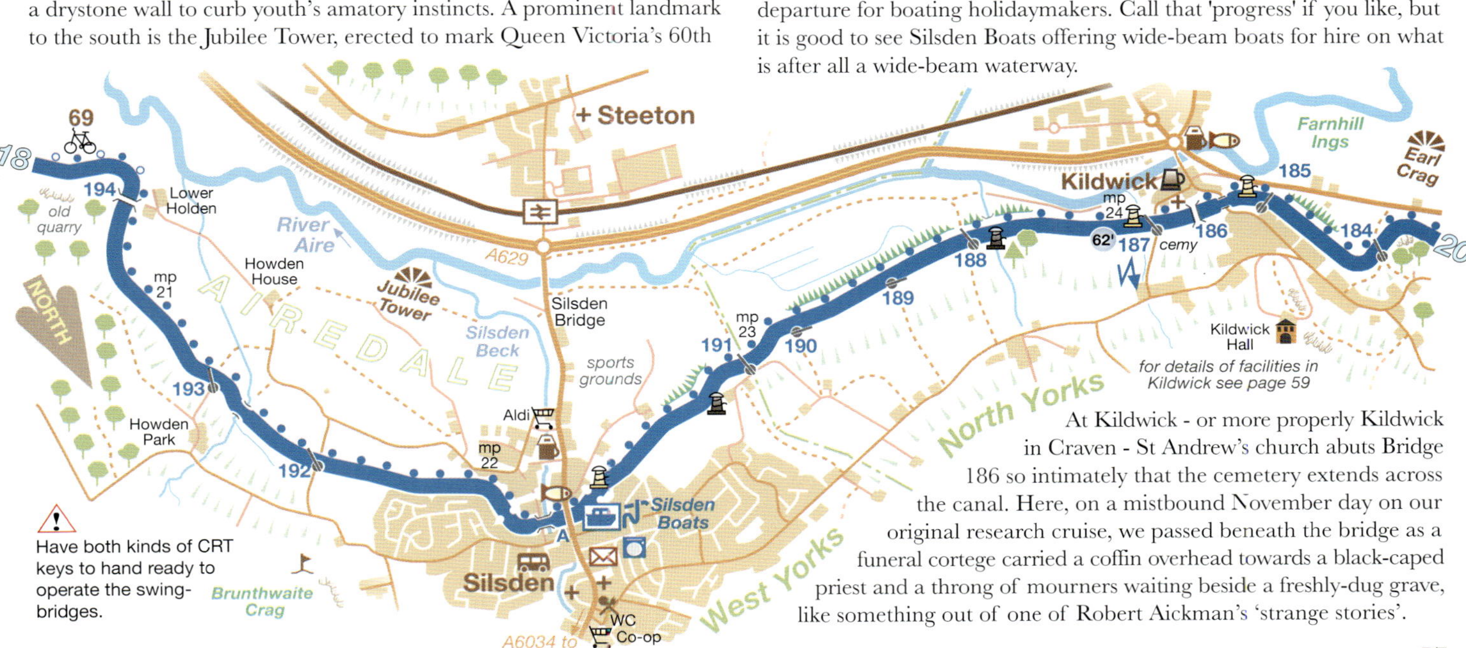

Have both kinds of CRT keys to hand ready to operate the swing-bridges.

At Kildwick - or more properly Kildwick in Craven - St Andrew's church abuts Bridge 186 so intimately that the cemetery extends across the canal. Here, on a mistbound November day on our original research cruise, we passed beneath the bridge as a funeral cortege carried a coffin overhead towards a black-caped priest and a throng of mourners waiting beside a freshly-dug grave, like something out of one of Robert Aickman's 'strange stories'.

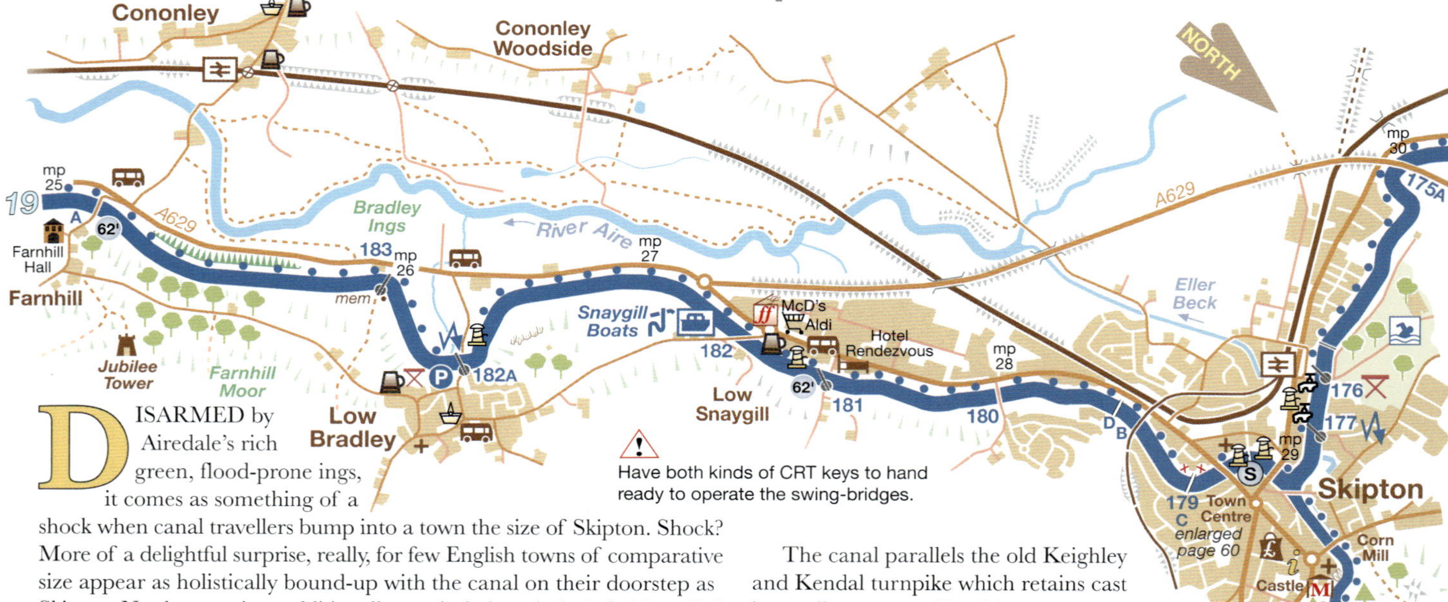

DISARMED by Airedale's rich green, flood-prone ings, it comes as something of a shock when canal travellers bump into a town the size of Skipton. Shock? More of a delightful surprise, really, for few English towns of comparative size appear as holistically bound-up with the canal on their doorstep as Skipton; Newbury springs additionally to mind, though don't feel impelled to write in with further examples.

The canal plunges into Farnhill Wood, a good, old-fashioned sort of arboreal glade made up of beech, birch, alder and sycamore growing sturdily out of a floor of moss and fern. Goodness, but for a lack of boats, this could be 'The Golly', it's *so* pretty. Scramble up through the bracken and heather to Queen Victoria's Jubilee Tower on Farnhill Moor for an invigorating view of the Aire Valley and its lines of communication. Across the valley, old smelting chimneys associated with quarry workings recall lost enterprises.

The canal parallels the old Keighley and Kendal turnpike which retains cast iron mileposts considerably more ornate than the Leeds & Liverpool's. By Bridge 183 a memorial recalls the death of seven Polish airmen hereabouts on 23rd September 1943. Their Wellington bomber was returning to Silloth on the Solway Firth when the port engine failed and they fell to earth. Briefly, the canal veers away from the main road to pass the time of day with an old mill village called Low Bradley.

Low Snaygill is home to another well-established hire base. Skipton's outlying industrial estates begin to jostle the canal, though it contrives to remain aloof until, beyond Bridge 180, terraced streets with cobbled alleys presage the town itself. Long trains of limestone rattle over Bridge 179B, 21st century descendants of the primitive tramway wagons which once brought the same commodity down to loading apparatus on the canal's Springs Branch. Between bridges 179B and 179D the ground floor of an old weaving shed has been reinvented as a drydock.

Skipton vies with Banbury in the proximity of its bus station to the canal. We like to think Pearson's users will relish the propinquity. Imagine our disappointment on learning that Pennine Motors, whose orange and grey vehicles we'd admired through four editions of *Pennine Waters*, had ceased operating. Never would helter-skelter bus rides to Barnoldswick or Settle seem so romantic or so death-defying again.

Boaters services are provided alongside the bus station by footbridge 179. Visitor moorings coincide with the point where short boats unloaded coal for the gas works. Who are we to mock the pretence that you've just arrived from Park Hill Staithe (Map 6) with a consignment to discharge, and that the minute you've shovelled and wheel-barrowed all fifty tons of

it into the engine house bunker, you'll be straight into Bizzie Lizzies for double fish & chips?

A larger than life statue of Freddie Trueman at full tilt welcomes you to the junction with the Springs Branch, and it's impossible not to be bowled over by Skipton's charming waterfront. Further visitor moorings are laid on between here and Bridge 177. Trip boats scurry hither and thither, an old tug dispenses ice creams, and coach loads of fish & chip devouring gongoozlers treat your trans-pennine conquests with suitable awe. Handsome textile mills, reinvented as luxury apartments, add gravitas.

Your boat has to be no longer than 35 feet if you want to explore the Springs Branch in a dignified manner, otherwise you'll have to reverse all the way back with your tiller between your legs. A far better bet is to board one of the regular trip boats which ply between the junction and the arm's mysterious terminus beneath the ramparts of Skipton Castle. Alternatively, do the journey on foot, popping into Stanforth's Celebrated Pie Establishment (which all but straddles the second bridge up) for sustenance en route.

Kildwick
Map 19

Picturesque settlement on the Bradford-Skipton turnpike known for its old bridge across the Aire and St Andrews, the 'Lang Kirk o' Craven'. In the graveyard lies John Laycock, a Victorian builder of organs, suitably remembered by a stone sculpture of one of his works.

Eating & Drinking
WHITE LION - Priest Bank Road (adjacent Bridge 187). Tel: 01535 636780. Comfortable inn serving food throughout from 11am (noon Suns) Ossett and Timothy Taylor ales etc. BD20 9BH

Low Bradley
Map 20

Eating & Drinking
SLATERS ARMS - Crag Lane. Tel: 01535 632179. Cosy whitewashed village pub reached via a nice footpath from Bridge 180. Food not served on Mondays or Tuesday evening. Local ales. BD20 9DE

Shopping
Enterprising village stores (Tel: 01535 634225 - BD20 9ES) 5 minutes walk from Swing-bridge 182A.

Skipton
Map 20

Styles itself, not without justification, as the 'Gateway to the Dales'. Indeed, one of those half-baked surveys voted Skipton the Best Place to Live in Britain, and it's easy to see why that might be the case. A market town of fourteen thousand souls, with grammar schools for both genders, a thriving economy with an eponymous building society at its heart, a cattle market (on the outskirts), shops to tempt the spendthrift in us all, high and low Anglican churches (in both senses of the term), a castle, base camp for the Settle & Carlisle railway, and, almost forgot, a canal. An unbeatable template for civilised urban living. Skipton Waterway Festival over the May Day Bank Holiday weekend.

Eating & Drinking
AAGRAH - Coach Street. Tel: 01756 790807. Housed in a former canal warehouse, this is one of a Yorkshire chain of Kashmiri restaurants. From 5pm. BD23 1LH
BEER ENGINE - Albert Street. Tel: 0737 578 9498. Micropub open from noon daily. BD23 1JD
BISTRO DES AMIS - Jerry Croft (High Street). Tel: 01756 797919. Delightful two-floor bistro presided over by Yorkshire/French patron. Meals served noon to 2.30pm and 5-9pm (noon to 6pm Sun). BD23 1DX
BIZZIE LIZZIES - Swadford Street. Tel: 01756 701131. Canalside fish & chip shop and restaurant. BD23 1QY
BOAT HOUSE BAR - Coach Street. Tel: 01756 701660. Cafe/bistro located within Pennine Cruisers' boatyard. Open 11am-10pm daily. BD23 1LH
LYCHEE - Keighley Road (Bridge 179A). Tel: 01756 798822. Canalside Chinese restaurant/take-away open daily (ex Mon) from 5pm (4.30 Sun). BD23 2LX

NARROW BOAT - Victoria Street (off Coach Street). Tel: 01756 797922. Splendid ale lovers' retreat from the hurly-burly of the High Street. Thoughtfully provided tasting notes set the tone and there's a good choice of food. Open from noon daily. BD23 1JE
RENDEZVOUS - Keighley Road (Snaygill). Tel: 01756 700100. Canalside hotel with conservatory restaurant and swimming pool open to non-residents. BD23 2TA
RHUBARB - Broughton Road. Tel: 01756 792781. Part of Herriots Hotel which backs on to the canal by swing-bridge 176. Breakfasts need to be booked by non-residents, but lunch and dinner are served from noon and 6pm respectively. BD23 1RT
TWO SISTERS - Mill Bridge. Tel: 01756 703712. Quirky bar/kitchen open 10am daily ex Tue. BD23 1NJ

Shopping

Some 'high street' casualties since we were last in Skipton, though where haven't there been! Craven Court Shopping Centre features the likes of Joules and Fatface but Lock-wheeler invariably makes a bee-line in the direction of three prize-winning pie makers: Stanforth's, Farmhouse Fare and Drake & Macefield. High Corn Mill (by Bridge 2 on the Springs Branch) houses an eclectic mix of shops and a 'working' water wheel. High Street hums with market stalls on Mon, Wed, Fri & Sat, an authentic street market of yore. The post office is on Swadford Street and there are Morrisons (Broughton Road) and Tesco (Craven Street) supermarkets within easy reach of the canal.

Things to Do

TOURIST INFORMATION - Town Hall, High Street. Tel: 01756 792809. BD23 1AH
BOAT TRIPS - Pennine Cruisers (Tel: 01756 795478) and Skipton Boat Trips (Tel: 01756 790829).
CRAVEN LEISURE - Aireville Park. Tel: 01756 792805. 25 metre indoor pool and cafe. BD23 1UD

CRAVEN MUSEUM & GALLERY - Town Hall, High Street. Tel: 01756 706407. Archaeology, costume, social history and travelling exhibitions. BD23 1AH
SKIPTON CASTLE - High Street. Tel: 01756 792442. One of the many castles which Cromwell 'knocked about a bit', (it endured a three year siege) but subsequently domesticised by the Cliffords and now open daily throughout the year to an admiring public. Tea room, shop and picnic area. BD23 1AW

Connections

BUSES - of use to towpathers, 'The Wizz' runs approx hourly daily to/from Barnoldswick, and service 580 to/from Gargrave & Settle. Tel: 0871 200 2233.
TRAINS - frequent Airedale Line trains to/from Leeds and Bradford. Also services over the scenic Settle & Carlisle and Lancaster routes. Tel: 0345 748 4950.
TAXIS - Station Taxis. Tel: 01756 700777.

Gargrave

Map 21

If Skipton is the Leeds & Liverpool's nicest town, Gargrave - picturesquely watered by the stripling Aire - surely qualifies as its loveliest village, and it has set out its stall to cater for visitors ever since the inception of the Pennine Way. Solely the interminable traffic on the A65 mars the village's idyllic existence on the edge of the dales. Iain Macleod (1913-1970), Chancellor of the Exchequer for just three weeks before his untimely death, is buried in the handsome church's graveyard.

Eating & Drinking

ANCHOR INN - Hellifield Road (adjoining Bridge 169A). Tel: 01756 749666. Cookhouse Pub with Premier Inn travel lodge accommodation. BD23 3NB
BOLLYWOOD COTTAGE - High Street. Tel: 01756 749252. Highly regarded Indian restaurant. BD23 3LX
DALESMAN CAFE - High Street. Tel: 01756 749250. Tea room and sweet emporium beloved of generations of walkers and cyclists. Closed Mon. BD23 3LX
THE FRYING YORKSHIREMAN - High Street. Tel: 01756 748345. Excellent fish & chip shop/rest. in the modern style. 11.30am-2.30pm and 4-8.30pm Mon-Fri and 11.30am-8.30pm Sat. Closed Sun. BD23 3RB
MASONS ARMS - Marton Road. Tel: 01756 749304. Cosy pub over the river. Accommodation. BD23 3NL

Shopping

Co-op (with cash machine) and pharmacy.

Connections

BUSES - Craven Connection service 580 runs to/from Skipton and Settle. Services 210/1 run two or three times daily to/from Malham, a delightful excursion. Tel: 0871 200 2233.
TRAINS - services to/from Skipton (handy for towpath walks), Leeds, Lancaster & Morecambe. Some Settle & Carlisle trains call too. Tempted? Tel: 0345 748 4950.

OYSTERCATCHERS pipe overhead as the canal traverses the Aire Valley overlooked by the russet-coloured crag of Sharp Haw. This is another of those benign stretches of the Leeds & Liverpool where you wonder why it isn't more popular with boaters, though recalcitrant swing-bridges (173 is a particular rotter) and cumbersome locks may offer some explanation. At least canallers are spared the headlong Gadarene rush of the A65, so busy that you suspect one half of West Yorkshire are on their way to the Lake District and the other half on their way back. The former Midland Railway's celebrated route from Leeds to Glasgow via Settle and Carlisle looks a much better bet as a means of civilised travel. Near Bridge 174, you catch a glimpse of Pendle Hill, a summit you'll become closer acquainted with the further west you go.

The offside tail-beam of Lock 30 bears a line of gilt-engraved poetry, though you may not necessarily feel poetically disposed towards its heavy gates. An aqueduct carries the L&L over Eshton Beck which feeds into the canal by milepost 33. The big works beyond the sports grounds on the banks of the river is a 'wound management' factory. £2 million had to be spent on repairing Eshton Road Lock in 2022. Refurbished now, for office use, an old canal warehouse overlooks Lock 31. Coal is still retailed canalside, but many years have passed since the warehouse would be stuffed with agricultural produce pending carriage out of the district by short boat. Three-day visitor moorings are provided above Higher Land Lock (No.32). Indeed, with time at your disposal, Gargrave makes an irresistible launch pad for Malham Tarn and Gordale Scar.

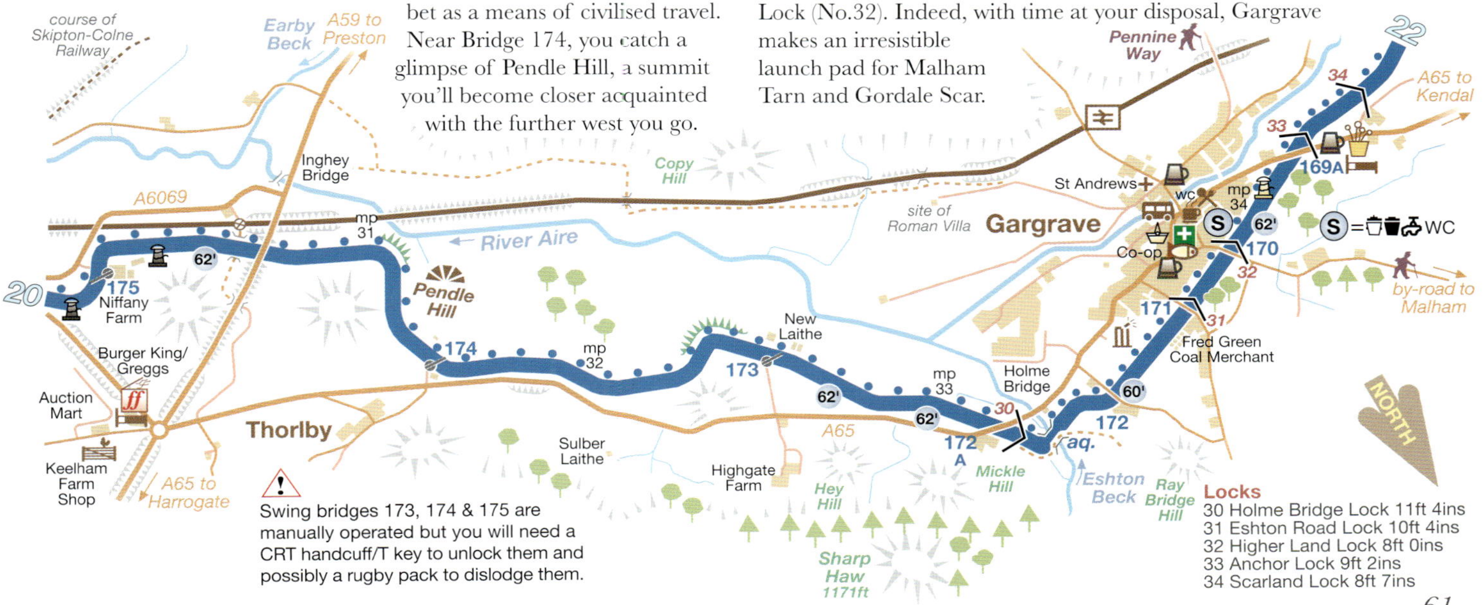

Swing bridges 173, 174 & 175 are manually operated but you will need a CRT handcuff/T key to unlock them and possibly a rugby pack to dislodge them.

Locks
30 Holme Bridge Lock 11ft 4ins
31 Eshton Road Lock 10ft 4ins
32 Higher Land Lock 8ft 0ins
33 Anchor Lock 9ft 2ins
34 Scarland Lock 8ft 7ins

PRIEST Holme Aqueduct carries the canal across the River Aire. Westbound, it's the last you'll see of this lovely river. It feels like losing an old friend. At the foot of Bank Newton Locks, the lintel over a cottage door reads "LLCC 1791", the year when construction of the canal recommenced westwards from Gargrave. The locks are notable in retaining clough (rhymes with 'Slough') ground paddles by their top gates, a refreshing survival flying in the face of uniformity, and so easy to operate compared with the laborious windings of a windlass. By No.37 there was once a busy maintenance yard, employed in the 1980s as a hire base.

Is there are lovelier pound on the whole of the inland waterways system than Marton Pool? Lock-wheeler has confessed elsewhere to being a serial adulterer where canalscapes are concerned. On another day, and in another Canal Companion, other lengths of canal might equally inspire him to hyperbole, but in favourable weather the five heavenly miles from Bank Newton Top to Greenberfield Bottom take some beating.

This is the Vale of Craven in Yorkshire's old West Riding, a celestial landscape characterised by what geologists call 'drumlins', and the rest of us, 'piddling hillocks'. The litter-bugging propensity of the Ice Age's retreating glaciers has left us a legacy of profound beauty, for these gentle summits are ravishingly Rubenesque in character. Furthermore, it would be churlish to argue that they had not been 'improved' by a man-made foundation garment of drystone walls, and agriculture hereabouts still appears to be husbanded along old fashioned lines. They were cutting the hay when we last passed, and whilst it was being mechanically bundled into black plastic rolls as opposed to romantic stooks, the scents were still a heady perfume rising up to the margins of the canal to mingle with those of meadowsweet and cranesbill. Fleetingly, in the vicinity of East Marton, canal travellers rub shoulders with back-packers on the Pennine Way. Their 256-mile tramp from Edale to Kirk Yetholm is all very laudable, but your 127 mile odyssey from Leeds to Liverpool surely deserves just as much kudos, and it's a toss-up which is more physically gruelling.

A DEPT at moving governmental goalposts, mere county boundaries present no obstacle to our political masters. In 1974 Ted Heath and his gang mugged the old West Riding town of Barnoldswick. By the time it had come to its senses it was, *fait accompli*, in Lancashire. Some of the locals still can't take it in, like crime victims who've never recovered their *sang-froid*.

Greenberfield Locks are couched in a sublime landscape. Three single chambers replaced a time-consuming staircase early in the nineteenth century, remains of which can still be seen, including a bridge-hole in a garden. By the top lock a feeder comes in from Winterburn Reservoir. Greenberfield was to have been the departure point for a branch canal proposed to link the main line with the market town of Settle. The mind boggles with the beauty of what might have been. A dwelling in Settle bears the name 'Liverpool House' in anticipation of the canal which never came.

Bordered by wastegrounds, 'Barlick's' designated visitor moorings do not do the town justice. By Bridge 153, Silent Night's bed factory occupies the site of Long Ing Mill. Ditto Esse Stoves, an old weaving shed. For many years the owners (proud would be pushing it) of an Esse, Lock-wheeler and his family learned to live with its periodic break-downs as one might make allowances for an unhinged relation. On such occasions the company would despatch a man called Granville to tinker with it - mend would be inaccurate. A delightful native of Accrington, and fount of East Lancs lore, now that it has been replaced by a reliable electric cooker, we miss Granville more than the stove.

But it is Rolls Royce who dominate the town's economy. Their Bankfield site borders the canal by Bridge 154A. Formerly the works was used for weaving: hence the saw-tooth north-light roofing. The first fighter jet engines were developed here toward the end of the Second World War. Nowadays the much extended plant makes components for famous jet engines such as the RB211: RB standing for Rolls-Barnoldswick.

continued overleaf:

continued from page 63:

The Rain Hall Rock Canal was cut to expedite the extraction of limestone from an aquatic, linear quarry face. In its half mile it included two short tunnels and a substantial viaduct carrying a by-road. Last used as long ago as 1893, most of its remains have been buried by landfill, though diligent exploration has its just rewards. Affectionately nicknamed 'The Spud', Barnoldswick lost its shuttle train service from Earby in 1965. Lower Park Marina is an important boating centre on a canal not widely endowed with such facilities. Shire Cruisers of Sowerby Bridge (Map 1) have an outstation here. That portable county boundary used to be between bridges 149 and 150, where a feeder from Whitemoor Reservoir enters the canal, overlooked by County Brook Mill with attractive employees houses alongside. The mill was still employing water power until the 1960s. Four generations of the Mitchell family have overseen a success story of fabric weaving - anything from deck chair covers and wind breaks to scarves, throws and curtains - and their substantial mill makes a handsome sight tucked into the folds of the landscape. Westwards, the land spills down from White Moor, before rising eastwards to Kelbrook Moor. The canal builders must have been grateful for a gap through which to cut their summit. Their leisure boating descendants are amply compensated for all the effort expended in getting here.

East Marton — Map 22

Heavenly stopover high up in the Dales with, appropriately enough, a lovely church with a Norman tower (Sunday service 10.45am) and charming interior with some exquisite stained glass, a pair of box pews, and commemorative tablets.

Eating & Drinking

ABBOT'S HARBOUR - adjacent Bridge 162. Tel: 01282 843207. Delightful cafe/restaurant also dealing in postcards, local books, confectionery etc. BD23 3LP
CROSS KEYS - iconic country pub sadly out of use as we went to press.

Barnoldswick — Map 23

Informally disposing with a syllable, the locals say 'Barlick', which makes it sound like a boiled sweet, as in: 'Pass round the Barlicks, Grandpa'. And truth be known, it *is* a hard exteriored, soft centred humbug of a little town whose older denizens still write West Riding of Yorkshire on self-addressed envelopes. A dozen local men perished when the troopship HMHS *Rohilla* hit a sand bar off Whitby in 1914.

Eating & Drinking

ANCHOR INN - Salterforth (adjoining Bridge 151). Tel: 01282 850055. Revitalized pub noted for the stalactites and stalagmites in the cellar! Food served Wed-Sun throughout from noon. BB18 5TT
ESSE - Long Ing Lane. Tel: 0771 994 2567. Canalside cafe adjacent Bridge 153 open daily ex Sun. BB18 6BJ
LOCK STOP CAFE - canalside Greenberfield. Fay & Tony's (plus helper Sue) welcoming kiosk opens 10.30am-4pm ex Fri. Pancakes, hot dogs, cakes and delicious ice cream for hot lock-wheelers. BB18 5SU
MARIO'S - Gisburn Road. Tel: 01282 816544. Lively Italian restaurant. Open for dinner (from 4/4.30pm) daily ex Tue & Wed. BB18 5HB

Shopping

Co-op on the site of the railway station. Post Office on Frank Street. As you have come to expect, Hutchinson's home made pie shop caught our eye. Launderette on Gisburn Road.

Things to Do

BANCROFT MILL - Gillians Lane. Tel: 0770 767 0177. Preserved compound stationary steam engine. Telephone or check internet for details of opening times and, in particular, steaming days. Refreshments and souvenirs available on such occasions. BB18 5QR

Connections

BUSES - of use to towpathers, 'The Wizz' runs approx hourly daily to/from Skipton. Mainline M5 runs half-hourly (hourly Sun) to/from Colne and Burnley via Foulridge and Nelson. Tel: 0871 200 2233.
TAXIS - Yellow Cars. Tel: 01282 814000.

Foulridge — Map 24

A deservedly popular overnight stop for boaters - in polite circles you say 'Foalridge' - the village's mills have for the most part found new uses and, along with terraced streets, evoke a workaday character.

Eating & Drinking

THE WHARF - Warehouse Lane. Tel: 01282 788200. Canalside cafe/bistro open 9.30am-5pm Mon-Thur and until 7pm Fri-Sun. BB8 7PP
FOULRIDGE HOUSE - Skipton Road. Tel: 01282 864146. Chinese take-away. Closed Mons. BB8 7PQ
FOUR ELEPHANTS - Skipton Road. Tel: 01282 864242. Indian restaurant open weekdays (ex Tue) from 4.30pm, 3.30pm Sat and Sun. BB8 7PY
NEW INN - Old Skipton Road. Tel: 0784 166 8987. Thwaites pub serving food from noon daily. BB8 7PD

Shopping

Ingham's, the butchers, do home made pies. Oddie's Bakery is one of sixteen sister outlets in East Lancs.

Connections

BUSES - as Barnoldswick.

METAMORPHOSIS comes to the cut. Depending on your direction of travel, you are about to exchange the wide open spaces of rural Pendle for the satanic textile towns of East Lancashire, or vice versa. But canals were built to generate trade. However pleasing to the eye, mile upon mile of picturesque bucolic wandering did not line promoters' pockets with dividends. As northerners were wont, pithily to put it: 'there's no brass wi' out muck'. Thus - in Pearson's never knowingly humble opinion - 21st century canal explorers are honour-bound to experience, and perhaps grow to savour, the muckier corners of the inland waterways system in order that it can be put into historical and geographical context. After all, don't New York's tour buses take you round The Bowery? Well they did when Lock-wheeler was parentally coerced to go there in 1967.

Foulridge is a magnet for the boater and the land-based canal enthusiast alike. A well-preserved stone warehouse (with Leeds & Liverpool Carrying Co brightly painted on its eastern gable end) adds dignity to the scene, and hosts a popular eating and drinking establishment. Once upon a time an unusual trestle bridge carried a railway across the canal. Services were ludicrously abandoned between the present railheads at Colne and Skipton

continued overleaf:

continued from page 65:

in 1970, but a vociferous group of stakeholders (SELRAP) aspire to having the eleven miles of track relaid and reopened. For, as they ironically point out, the journey by rail between Colne and Skipton now distances a hundred miles, via Burnley and Leeds, and takes a couple of hours. Part of the line's trackbed can be explored on foot in a north-easterly direction. Another notable feature is a restored limekiln erected early in the construction of the canal to provide lime for mortar, which relied initially on Lothersdale limestone brought in by 'lime gals'. Not, to some of our crew's regret, stocky local wenches with descendants in the vicinity, but sturdy Galloway ponies.

Towpath-less, and confined to one-way working, Foulridge Tunnel was five long years in the making, the cut & cover technique being applied. Leggers gave way to steam tugs in 1880, which in turn survived until 1937, by which time trade over the summit was negligible and largely diesel-powered in any case. Traffic-light signals control the entry of boats through the tunnel, and a passage should take around quarter of an hour. The 1,640 yards long tunnel's claim to fame lies not so much in the fact that it is the longest on the Leeds & Liverpool Canal, but that it was the scene, in 1912, of one of the more bizarre events in the annals of bovine history, when a cow called Buttercup fell into the canal at the western end of the tunnel and swam through to the eastern end before being rescued and revived with brandy. The local tourist authority are missing a trick in not re-enacting Buttercup's misadventure annually.

Barrowford Locks lower the canal some seventy feet down from its six

mile summit pound, four hundred and eighty-seven feet above sea level. The flight derives its name from a former weaving village to the west, whilst the important town of Colne straddles a hillside in the opposite direction, the clocktower of its Town Hall dominating the view against the backdrop of the mid-Pennine moors. Colne's most famous son was Wallace Hartley, the heroic bandmaster on the *Titanic* who phlegmatically continued to conduct *Nearer, My God, To Thee* as the liner sank beneath the icy waters. His appropriately decorated grave can be found in the town's cemetery on Keighley Road.

Back alongside the canal, Barrowford Reservoir receives what surplus water the summit might have to offer. A small car park is provided for motorised visitors to the flight. Blacko Tower overlooks proceedings. It was erected in 1890 by a grocer called Jonathan Stansfield who preferred follies to women and alcohol. His loss.

Welcoming you back to reality, the M65 spans the foot of the flight, and an aqueduct carries the canal over Colne Water which flows down off the Forest of Trawden. The towpath changes sides at Bridge 142 as urbanisation begins to make its presence felt. Bridge 141A is overlooked by a fine specimen of a canal warehouse, a brick-built cousin of the structures at Burnley and Blackburn which has found a sustainable new role for itself as part of a health centre. There was once a corresponding warehouse on the south side of the bridge and, opposite that, a skating rink of all things. Old maps depict the presence of several canalside mills; all vanished, most recently the Pendle Street weaving shed, a wasteground now, symbolically pining for redevelopment.

UNIMPEDED by locks or swing-bridges, boaters can give their full attention to the flamboyant mill architecture of East Lancashire. It's just a shame not many examples are engaged in textile production now, though William Reed & Co's Spring Bank Mills are a notable exception, whilst cycling enthusiasts will be tickled to find they're passing the premises of Carradice, saddle bag makers since 1932.

'Brierfield is not noticeable as an entity', was Nikolaus Pevsner's opinion, master of the architectural put down. St Luke's since disused church came in for further scorn: 'It is buildings like this that make

of rather less brio than Salts (Map 17) - whilst Brierfield Mill itself has been re-invented as a £32 million mixed use redevelopment scheme entitled "North-Light."

Bridge 137 is called Lob Lane Bridge, in apparent homage to the sport of road bowling or cobble hurling, much enjoyed hereabouts in the 19th century. On its south side, a former weaving shed

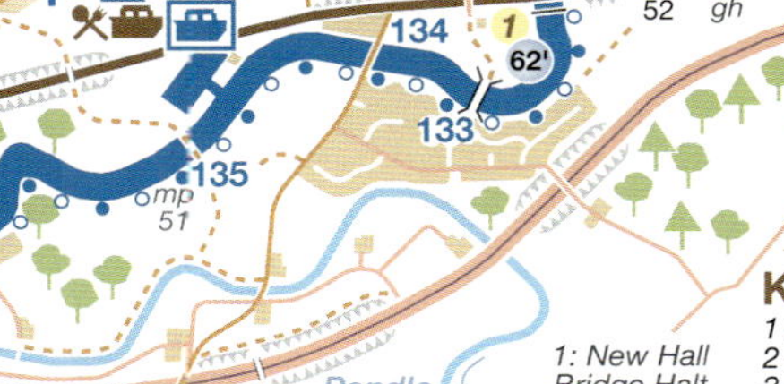

people refuse to take Victorian architecture seriously'. Perhaps he'd had a particularly bad picnic lunch that day, for guidebook compilers are notoriously subjective. But Brierfield's mills can hold their own with any along the canal corridor, even if Hollins Mill has become a retail outlet - though

has been fetchingly transformed into smart apartments.

Good honest countryside intervenes, and a quality of remoteness, heightened by the proximity of urbanisation. Pendle Hill (1831ft) looks good from this angle, and the villages

cuddled in the folds of its south-facing flank hold memories of the notorious Lancashire Witches. Reedley Marina reflects the canal's change of use from work to leisure. Up until the 1930s, an electric tramway linked Burnley, Nelson and Colne, notable in that a short section through

continued overleaf:

for details of facilities in Burnley turn to page 69

continued from page 67:

Reedley Hallows was uniquely owned by the parish council. Bridge 134 carries Barden Lane across the canal. Barrett's Directory of 1911 lists its occupants engaged in a roll-call of lost trades: carter, collier, clogger, shunter, weaver, boatman and tripe dealer. At railway bridge (132A) the canal widens at the site of Reedley Colliery loading basin.

Overlooked by an old gasometer, the canal feels its way into Burnley, tentatively at first, skirting a pretty municipal park as if to gain confidence. An aqueduct carries it over the River Brun, one of the shortest rivers we can think of, rising on the flanks of Worsthorne Moor, just three or four miles to the east, and colliding with the Calder in some obscure corner of Burnley town centre. In earlier times a mineral railway serving Bank Hall Colliery passed beneath the aqueduct. The mine's barge loading basin is still in water by the sharp turn in the canal, as is a drydock nearby. Bank Hall's first shaft was sunk in 1860, but perhaps its busiest period for canal traffic was in the middle of the twentieth century when it was the principal supplier of coal to Whitebirk power station at Blackburn. The pit was finally closed in 1972; on safety grounds, they said.

Burnley's most astonishing canal gesture, and one of Robert Aickman's *Seven Wonders of the Waterways*, is the embankment which carries the Leeds & Liverpool across the broad, converging valleys of the Calder and Brun, to the east of the town centre. Sixty feet high, and three-quarters of a mile long, Burnley Embankment affords the canal traveller a bird's eye view of the town, a view dominated by the copper dome of the Town Hall whose clock will tell you accurately enough the time of day, if not what century you're in. But in truth, from a boat, or from the towpath, the embankment - significant an engineering feat of its era as it is - lacks perspective. You get a better view from the bus station, in amongst frequent gear-crashing departures for outlandish destinations like Stoops, Bacup and Harle Syke; the latter sounding like a tyrant in a Gothic novel.

Turf Moor stadium is the home of Burnley, a football club with a pedigree as proud as Blackburn's down the cut. The first 'derby' between the two clubs in 1888 resulted in a 7-1 win for Blackburn, whose star player's day job was as a violinist in the Halle Orchestra. The slate rooftops and glazed chimney pots of densely packed terraced streets - which gleam dully after a shower like tarnished silver - keep company with the embankment's rim, whilst the horizon is a wild backdrop of moorland; an escape clause for all those consigned to an occluded existence in these Pennine textile towns. On one of the tops to the south-east, Coal Clough wind farm gesticulates madly as if trying to draw our attention to something: an unexpected item in the bagging area perhaps.

After many years of lugubrious decay - we still shudder at the memory of mooring here one freezing night when the wind was blowing unchecked off the Pennines - Finsley Gate Wharf has belatedly had a makeover by the Canal & River Trust with the help of Heritage Lottery funding. In marked contrast, on the occasion of our visit to update this edition a guitarist was suavely entertaining cocktail-sipping patrons of the wharf's new restaurant. We spotted a kingfisher patrolling its low-flying beat just here where the canal begins to twist and turn through a district once stuffed to the gills with the spinning mills and weaving sheds which kick-started Burnley's 19th century prosperity. Now the canal corridor - branded 'On the Banks' - has a role in 21st century regeneration as part of the town's 'Knowledge Quarter', a campus of the University of Central Lancashire, a new lease of life which the weavers of the past are unlikely to have forseen.

Manchester Road climbs up from Burnley's Town Hall and celebrated Mechanics Institute to Burnley Wharf with its imposing warehouses and overhanging canopy: a fitting location for the Weavers' Triangle Visitor Centre housed in the wharfmaster's house and toll office. Of unique interest is Slater Terrace by Bridge 130, where ground floor workshops were topped by a terrace of two-storey workers' cottages: an early example of 'working from home'.

An aqueduct carries you across the M65 then, just as you were beginning to think that Burnley would go on forever, Gannow Tunnel swallows you up and spits you out in a suburb which might have been named after a mill girl. Gannow's 559 yards long, towpath-less bore provokes walkers and cyclists into a detour.

Barrowford

Time was when Lock-wheeler and his polytechnic pals would stop here for fish & chips on their weary way back to Manchester after a full day up in the dales around Malham. Still a textile community then - albeit one of handsome handloom weavers' cottages - now it's been exfoliated clean and turned into a bijou village of soaring property prices and snarled up traffic. And, give him his due, Lock-wheeler still likes it immensely. A 10 minute walk along the pavemented B6247 will get you there; better still an off-road route from Bridge 142 leads across Colne Water and under the motorway to the centre of the village.

Eating & Drinking

BANKERS DRAFT - Gisburn Road. Tel: 0773 987 0880. Micropub. Closed Mon & Tue. BB9 6HQ
GEORGE & DRAGON - Gisburn Road. Tel: 01282 618710. Local opposite heritage centre. BB9 6JD

Shopping

Niche shopping has muscled out the old corner shops, and blue ribboned Booths have opened a supermarket, why even David Beckham has been seen to shop in Barrowford's boutiques. Farmers' market 2nd Sunday in the month. The old Higherford Mill houses an artist studio collective open to the public Wed-Sun 10am-4pm.

Things to Do

PENDLE HERITAGE CENTRE - Park Hill. Tel: 01282 677150. 17th century farm buildings once owned by the Bannisters from whom Sir Roger, the first 4 minute miler is descended. Pendle Witches. Open daily 11am-4pm. Small admission charge for museum and walled garden. Shop & excellent cafe. BB9 6JQ

Connections

BUSES - Burnley Bus Co. service 2 half-hourly links with Nelson (hourly Sun). Tel: 0871 200 2233.

Colne

Friendly, characterful and historic hilltop town most easily reached by bus from Foulridge or Nelson.

Eating & Drinking

BANNY'S - Vivary Way. Tel: 01282 856220. Coach party popular fish & chip restaurant and take-away adjoining Boundary Mill shopping outlet. BB8 9NW

Shopping

Asda, Lidl and Sainsbury's supermarkets. Market Hall.

Connections

BUSES - see Foulridge and Nelson.
TRAINS - hourly (bi-hourly Sun) Northern service to/from Nelson, Brierfield, Burnley, Rose Grove, Blackburn, Preston and Blackpool. Tel: 0345 748 4950.
TAXIS - Union Cabs. Tel: 01282 870200.

Nelson

You'd need to be one-eyed not to take to Nelson, not so much a different country as a different continent.

Eating & Drinking

SPICE OF INDIA - Manchester Road. Tel: 01282 613353. Deservedly popular Indian restaurant open from 5pm daily (2pm Sun). Take-aways too. BB9 7HD

Shopping

Canalside Morrisons. Mellins butchers, where we used to buy black pudding, has become the Pendle Halal Meat & Poultry Centre. Mouthwatering Asian sweet shops. Indoor market Mon-Sat. Asia Bazaar.

Things to Do

PENDLE WAVELENGTHS - Leeds Road. Tel: 01282 661717. Sloping beach, wave machine and black hole water slide. Got your cossie on underneath? BB9 9TD

Connections

BUSES - Burnley Bus Co. M3/4/5 at frequent intervals to/from Burnley and Colne. Tel: 0871 200 2233.
TRAINS - as Colne.

Burnley

It's the dialect that delights, and to gain a sense of how Burnley once saw itself, pass through the revolving doors of the library and gaze in awe at the glass ceiling. Thus was civic pride demonstrated in 1930. In those days glass ceilings didn't thwart, they propelled you to the stars, and where better to launch yourself than in a public library?

Eating & Drinking

BRIDGE BIER HUIS - Bank Parade. Tel: 01282 411304. Free house offering a wide choice of micro-brewery ales, open from noon Wed-Sun. BB11 1UH
FINSLEY GATE WHARF - Finsley Gate (Bridge 130E). Tel: 01282 940480. New cafe/restaurant regenerated from derelict canalside wharf. Open 10am-5pm Mon-Wed; 10am-11pm Fri & Sat; 10am-8pm Sun. BB11 2FG
THE PALAZZO - Grimshaw Street. Tel: 01282 902123. Italian restaurant open from noon daily. BB11 2AS

Shopping

Sainsbury's and Tesco supermarkets within walking distance of the canal, but far more atmospheric is the 1960s first floor Market Hall. Try Haffner's pies. Carnivores can't fail to be impressed by the size and scope of Butchers Fayre on Queen's Lancashire Way.

Things to Do

WEAVERS' TRIANGLE - Burnley Wharf, Manchester Road. Tel: 01282 452403. History centre with access on selected dates to Oak Mount Mill Engine House. Open April to September Sat-Tue pm. BB11 1JZ

Connections

BUSES - services throughout East Lancs. Catch No.8 for a moorland ride to Bacup. Tel: 0871 200 2233.
TRAINS - Central and Barracks stations for Colne and Blackburn etc; Manchester Road for Manchester, Leeds and Preston. Tel: 0345 748 4950.
TAXIS - Kings. Tel: 01282 422551.

YOU can picture Rose Grove: clip-clopping down the cobble-stones in her headscarf, clogs and second-best coat; day-dreaming of the lad who'd taken her to the pictures the previous night; impatient to spill all the lurid beans to the lasses at neighbouring looms. Perhaps he was a miner. Perhaps he worked at the iron works. Perhaps he was a 'passed cleaner' at the engine sheds. Long buried beneath the M65, '10F' was one of the last three motive power depots retained to service steam until August 1968. Taking their water from the canal, it was said that you could always tell a Rose Grove tender when it went to the works by the shoals of minnows inside.

Nowadays, head-scarfed mill gills in the Harry Rutherford mould are conspicuously absent, but Rose Grove is a good spot for a boating breather, and the Canal & River Trust provide secure visitor moorings at their offside maintenance yard by Bridge 126A. A stone warehouse sets the tone and there's a good range of boating facilities.

Between Rose Grove and Hapton the canal was diverted to facilitate construction of the motorway. Hapton chemical works closed in 2006. Opened in 1846, it was reputedly the largest in the north of England, but the Riley family, who owned it, had to pay substantial death duties at the end of the First World War (as a result of fatalities within the family)

and were forced to sell out to Blythe's at Church (Map 27). Following demolition, houses have been built on the site, and a nature trail developed on an adjoining spoil tip. Down in the valley, Padiham's prominent Victorian church has the perfect backdrop foil of Pendle Hill. Gone, though, are the cooling towers of Padiham power station, demolished following closure in 1993, ditto those across the canal at Huncoat; a once commonplace but now vanishing aspect of the British landscape. Skirting Hapton, the canal arcs to cross a series of ravines, or cloughs (as 'enough', not 'ow!') as they call them hereabouts, gouged out by water-courses cascading down to the Calder. Hapton Castle had its origins in the fourteenth century but was ruinous by the eighteenth. Remnants can be discerned amidst limestone rock formations in Castle Clough Wood. There is much for the diligent industrial archaeologist to uncover as well.

West of Hapton, green fields lap the margins

Key 1

1 former baths
2 sites of textile mills
3 sites of iron works
4 site of Rose Grove MPD
5 site of Hapton Valley Coly.
6 site of chemical works
7 former dyeworks
8 site of brickworks
9 site of vitriol works

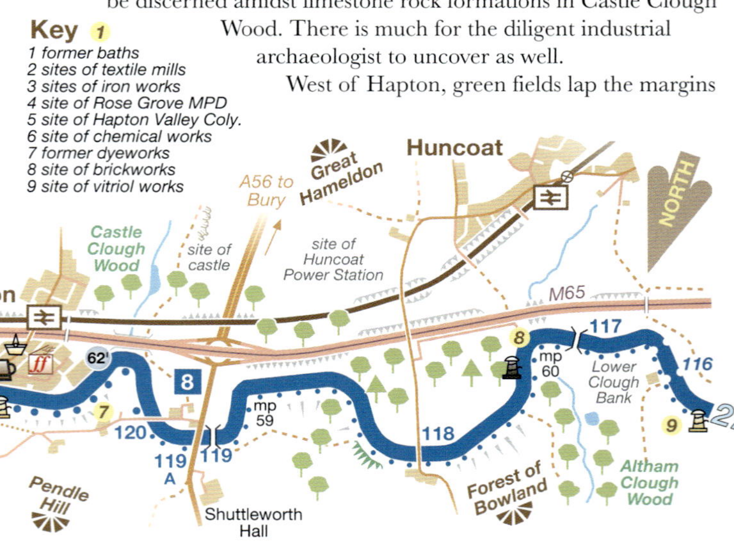

of the cut, stretching from the towpath's drystone wall boundary to a horizon of moorlands, the Calder Valley, Trough of Bowland and Pendle Hill. Savour them, for there is an inevitability about the onslaught of Burnley and Blackburn. And ignore the middle distance, where some philistines have seen fit to plonk a distribution hub; as if there were not enough brownfield sites in the district where mills once stood.

Southwards, the quarry-scalloped edge of Great Hameldon (1343ft) prefaces the lonely tops of the Forest of Rossendale. But the motorway effectively curtails the proximity of this landscape, and it is northwards, over the valley of the Calder (no relation to the navigable West Riding namesake, though born of the same watershed) that the canal traveller's eye tends to be drawn. You may even catch glimpses of Penyghent in the wide blue yonder. In the middle distance the parkland demesnes of Read, Simonstone and Huntroyde contrast with stone-walled pastures patterned by darker masses of woodland, ascending to the mottled flanks of Pendleton Moor.

Rose Grove — Map 26

With pub names like The Railway and The Junction, Rose Grove's railway heritage is manifest, its canal credentials less so, though it's odd to think that this characterful little East Lancs community lies nearer Leeds than Liverpool. In place of iron works and cotton mills the local economy rests on the shoulders of a business park. Gannow public swimming baths closed in 2004 and have become a pentecostal church.

Eating & Drinking
SYCAMORE FARM - Liverpool Road. Tel: 01282 427101. Greene King Farmhouse Inn pub/restaurant serving food from 8am daily. BB12 6HH *Fish & chips either side of Bridge 126A, plus several other take-aways, make this a good spot for an overnight stop.*

Shopping
Londis convenience store, pharmacy, bakery and post office.

Connections
TRAINS - Northern services operate hourly (bi-hourly Sun) to/from Burnley, Nelson and Colne, and Blackburn, Preston and Blackpool. Tel: 0345 748 4950.

Clayton Le Moors — Map 27

A little town of two halves, inhabitants being referred to as 'top-enders' or 'bottom-enders' depending on which side of Whalley Road canal bridge (114B) they live: reputedly, they rarely mix! Clayton Le Moors Harriers were founded in 1922. One of their best known members was the fell runner Alan Heaton (1928-2019) who conquered all 214 'Wainwrights' at the age of 57. When, in his late seventies, the hills became too much for him, he took to running along the Leeds & Liverpool Canal's towpath instead.

Eating & Drinking
OLD ENGLAND FOREVER - Church Street. Tel: 01254 383332. Sympathetically refurbished back street pub open from noon daily and featuring Bank Top Brewery ales from Bolton. BB5 5HT

Shopping
Co-op and convenience store, post office (with cash machine), pharmacy and sandwich shop. Olivia Ryan offers home made deli meals to take-away.

Connections
BUSES - 'Hotline' service 152 runs half-hourly between Burnley and Blackburn (hourly Sun), service 6 operates quarter-hourly (every 20 mins Sun) to/from Blackburn via Accrington, an interesting town to explore.

Church — Map 27

AKA Church Kirk - just in case you're in any doubt as to its ecclesiastical origins - is a parish with a fascinating past. A hotel since the late 1940s, Dukenhalgh Hall can trace its origins (like St James' itself) back to the 13th century and naturally claims a resident ghost, an 18th century governess spurned by a cad. Long lost industries include a naphtha distillery, mordant works and dyeworks specialising in Turkey Red. Today's large gifts and greetings card factory dulls by comparison.

Eating & Drinking
MONTE CRISTO - Henry Street. Tel: 01254 238400. Italian restaurant open Tue-Sat 5.30-9.45pm, Sun 4-8.45pm. BB5 4EP

Connections
BUSES - services 6/7 operate frequently, daily from a stop close to Bridge 112 to Accrington. TRAINS - as Rose Grove.

Rishton — Map 27

The broadcaster and horticulturalist, Christine Walkden, (*Gardeners' Question Time* etc) hails from here. The imposing Victorian church is by Maycock & Bell.

Eating & Drinking
BALTI HOUSE - High Street. Tel: 01254 883301. Indian restaurant and take-away. BB1 4JZ
Also a canalside cafe, two pubs, and fish & chips.

Shopping
Co-op, post office, pharmacy, bakery, and butcher.

Connections
TRAINS - as Rose Grove.

27 LEEDS & LIVERPOOL CANAL Church & Rishton 8mls/0lks/3sbs/3hrs

YOU get an even suntan - subject to meteorological terms & conditions - as the canal contrives to double the distance between Clayton and Blackburn, wriggling with the contours. In cargo-carrying days this profligacy with the miles may have irked, but now we are grateful for as much of this invigorating and stimulating canal journey as we can get; valuing its long-windedness, letting the M65 motorway cater for the speed merchants, who wouldn't know a topographical nuance if they ran one over.

Suburbia heralds Clayton Le Moors, a community not quite so bucolic as its name implies. The comedian, Eric Morecambe, worked as a Bevan Boy at Moorfield Colliery during the Second World War. Not so amusingly, the mine was the scene of an undergound explosion in 1883, causing the deaths of sixty-eight workers including boys as young as ten. We looked in vain for Appleby's flour mill, operators of a fleet of barges, and the soap factory which specialised in the production of 'floating soap'. The towpath swaps sides at Enfield Wharf which marked an hiatus in the canal's westbound construction for several years while more capital was raised. Despite disreputable appearances, it plays host to Accrington Sea Scouts and its warehouses wait - probably hopelessly - for a new use. This is cricket country! No, not the soft, self-centred poetical village cricket of Edmund Blunden's eponymous book, but the raw stuff of the Lancashire League, a semi-professional circuit of surprising intensity which has featured some of world cricket's most celebrated players down the years. Rishton, for example, can boast of having Viv Richards, Michael Holding and Allan Donald on its books. Enfield's imposing ground lies up on the hillside to the east. Whin Isle Farm lies in splendid isolation between the canal and the motorway, as effectively remote as a croft in Wester Ross. Overlooked by St James's church, a replica milepost marks the halfway point - sixty-three and five-eighths miles -

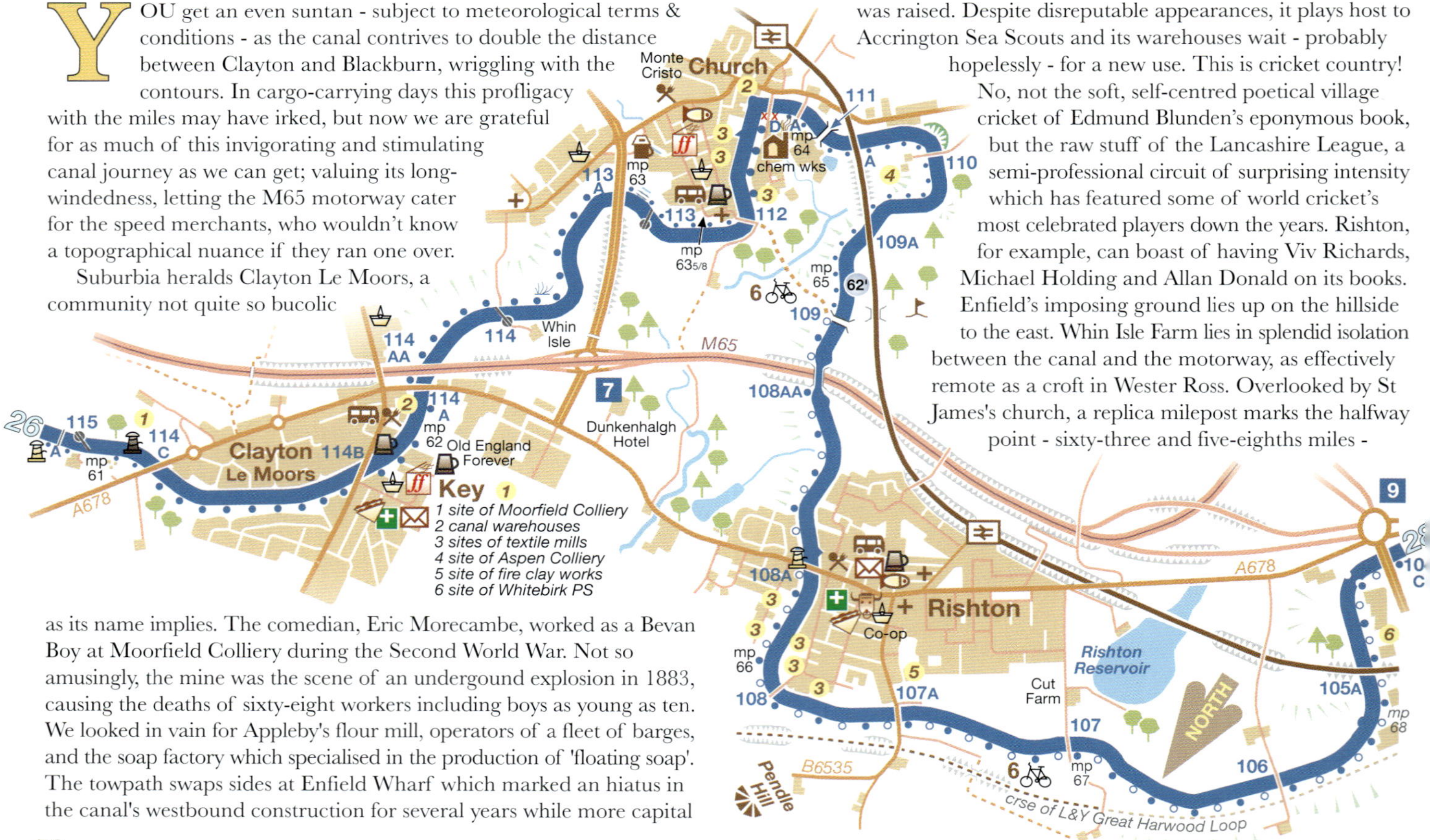

between Leeds and Liverpool. There were proposals, towards the end of the 19th century, to extend the Peel Arm (dug initially to serve a calico works) into the centre of Accrington, but sadly (for Accrington is a fascinating place) they never came to fruition. A substantial, three-storey canal warehouse overlooks a right-angle bend in the canal at Church Wharf. Flyboats once unloaded cotton here, and despite the ruined state of the building, still seem almost tangible.

Once there were fifteen mines in the neighbourhood of Church and Oswaldtwistle. Chemical production is another local industry. William Blythe's are still flourishing a hundred and fifty years after a canny Scot from Kirkcaldy eponymously set-up in business, attracted by the canal's transport potential. Prominent notices advise boaters to remain inside their craft in the (unlikely) event of chemical leakages signalled by sirens. Walkers and cyclists are, by implication, expendable.

Take to the path which swoops down into the Aspen valley between bridges 112 and 109 and you can beat the boat with ease. Now carried on an embankment, when the East Lancs Railway was built in 1848, the line crossed the valley on a lengthy timber viaduct. It took twenty years of spoil-tipping to bury the bridge. Aspen Colliery opened in 1869. Coke was produced in a series of 'beehive' ovens (nicknamed the 'fairy caves'), traces of which remain beside a reed-filled basin adjacent to the railway.

Contemporaneously androgynous, the landscape combines the feminine allure of pastures backed by waves of moorland, with the masculinity of

industry and urbanisation. Chameleon-like, the canal seems to alter its character and sexual persuasion in response to its environment. Rishton is imbued with XY chromosomes, and though its gaunt mills have tumbled, tightly-packed terraced streets remain. Rishton Reservoir was built in 1828 to augment water supplies to the western end of the canal. The Burnley - Blackburn railway crosses its southern end by causeway. The old loop line to Great Harwood closed to passenger trains in 1957. Canon Roger Lloyd, one of many clergymen with a deep interest in railways, wrote of it in *Farewell to Steam*, recalling that when he was vicar at Great Harwood in the 1920s, 75% of the working population were unemployed.

Whitebirk Power Station was the source of the last regular trade on this part of the canal, until the thirteen week freeze of '63 finally convinced the Central Electricity Generating Board of the unreliability of canal transport. The generating plant had been built in 1921. Coal was discharged from barges on the offside - a wasteground of saplings now - and conveyed overhead into the plant on the towpath side of the canal. The bulk of Whitebirk's coal came from Bank Hall Colliery at Burnley (a lock-less fifteen miles to the east), but a proportion also originated from pits in the Wigan coalfield. They stopped making sparks fly here in 1976 and the power station was demolished seven years later, to be replaced by a retail park which looks as if its days are numbered too. Tug-of-war competitions used to be held across the canal where it narrows at the site of an abandoned swing-bridge. The losers, presumably, got wet!

Bridge 112, Church

BLACKBURN impinges its urban - though not necessarily urbane - personality on the canal, obliterating all sense of countryside for the best part of five miles. There are still, however, green horizons to savour: Jubilee Tower on Darwen Hill to the south; the woodlands of Witton Country Park to the north. The outskirts let you in lightly; allotments and bowling greens leaven cobbled streets and ginnels. But this is primarily a part of the journey which should be devoted to disinterring the remains of King Cotton; though it must be emphasised that the local authority have not been content to live in the past, and a good deal of work has been undertaken to revitalise Blackburn's canal corridor: it just needs

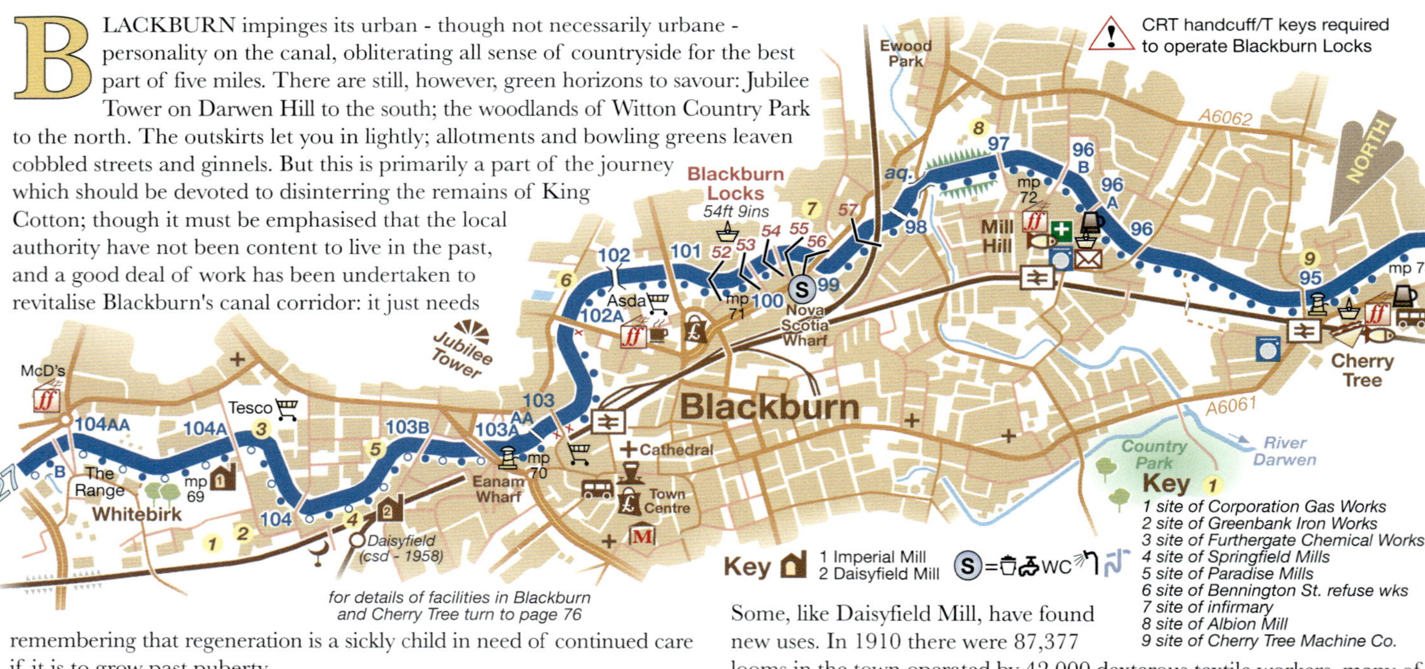

for details of facilities in Blackburn
and Cherry Tree turn to page 76

remembering that regeneration is a sickly child in need of continued care if it is to grow past puberty.

Every baby boomer remembers from *Sgt. Pepper* that there were "Four thousand holes in Blackburn, Lancashire". But once there were two hundred factory chimneys as well, one of them - at 312ft, probably the highest in the UK - stood canalside at Bennington Street refuse works, built in 1888 of nearby Accrington's famously durable bricks. Durable, but dynamitable, and now you can count Blackburn's post industrial smoke stacks on your fingers. Many mills have shuffled off their mortal coil.

Some, like Daisyfield Mill, have found new uses. In 1910 there were 87,377 looms in the town operated by 42,000 dexterous textile workers, many of them children as young as twelve. Between bridges 104A and 103A the canal snakes its way between pigeon-haunted, buddleia-blossoming walls whose purpose now is impossible to penetrate. The canalside presence of Webbox pet foods, and Graham & Brown - who've altruistically decorated a bridge-hole with sample wallcoverings - illustrates, however, that business can still thrive here, even if the raw materials no longer arrive by boat. Across the railway from Bridge 104 a large mosque impacts on the urban landscape in an echo of the mills which preceded it. Imperial Mill, a vast, sprawling, redbrick, double-domed dinosaur hanging on for dear life was designed by the renowned mill architect, Sidney Stott. Erected in 1900, it ceased spinning eighty years later, but something is still going on in there, judging by the whirr of lathes within.

Eanam Wharf was one of the Leeds & Liverpool's most significant goods depots. Twenty-five thousand natives gathered to see its opening in 1810. The local rag's 'shipping list' for the 27th June of that year quotes the arrival of the barges *Dispatch* carrying yarn, molasses and tallow; *Defiance* with timber and lead; and *Ten Sisters* bearing malt and earthenware. A hundred and forty-five years later John Seymour arrived here aboard his converted Dutch sailing barge *Jenny the Third* (*Sailing Through England*, Eyre & Spottiswoode 1956) and found the depot stuffed with cotton bales brought by lorry from Liverpool docks. For short boats, wide boats - call them what you will - the writing was already on the wall. Eanam Wharf's substantial warehouse now houses a business centre and a Caribbean restaurant: clearly, giants no longer roam the earth.

Providing a panoramic view of the town - dominated by the Town Hall's 1960s addition - the canal zigzags around the southern fringe of the centre. A flight of six locks (tediously equipped with handcuffs, but recently given a welcome lick of paint) carries the canal down from its elevated position on the western edge of town to Nova Scotia Wharf. Crossing an embankment, Ewood Park football stadium appears above humble terraced rooftops. In common with many Lancashire clubs, Blackburn Rovers' history goes back to the dawn of professional football. They were First Division champions twice in the years leading up to the First World War, and FA Cup winners on six occasions. After a period in the footballing doldrums, they came to prominence again in 1995, winning the Premiership under the management of Kenny Dalglish with a team which included Alan Shearer, built from millions invested by the local steel magnate Jack Walker. A succession of managers have found it hard to replicate that success since those heady days. and Blackburn have been plying their trade in the Championship since 2012.

Blackburn
Map 28

This is the town which Alfred Wainwright, the peerless fell-walker and guide book compiler, was desperate to escape from as a young man. Some canal travellers may yet sympathise, but we have always warmed to Blackburn's proud Victorian and Edwardian architecture, still recognisable - here and there - as the townscape the early film-makers, Sagar Mitchell and James Kenyon, had ground-breakingly pointed their lens at. Cloth caps, shawls, clogs and trams (with which the borough persisted until 1949) may have given way to taqiyah, achkan, khussa and buses (an impressive new bus station opened in 2015) but considerable municipal pride was being focussed on the Cathedral Quarter development when we visited Blackburn to research this guide. Covent Garden panache was promised and Britain's first new 'cloister garden' in six hundred years. Reacquainting ourselves with the town (it failed in attempts to become a city in 1994 and 2022) for the purposes of this update, we were incredulous to discover that Waves, the state of the art swimming centre opened in 1986, had been demolished and replaced by a cinema. So much for encouraging the public to take exercise! But to end on a more positive note, Thwaites, whose brewery tower had been such a landmark in the town until it was razed to the ground in 2019 - and who had controversially sold some of their best loved beers to Marston's - have moved to new premises at Mellor Brook, five miles to the north-west.

Eating & Drinking
AL CRISTALLO - Cathedral Square. Tel: 01254 447234. Italian restaurant. BB1 1FB
CAFE AT BLACKBURN CATHEDRAL - Cathedral Square. Tel: 01254 260520. Cafe in cathedral. Open Mon-Sat 8.30am-4.30pm Tue-Sat. BB1 1FB
CALYPSO CARIBBEAN - Eanam Wharf. Tel: 01254 261020. West Indian cuisine alongside the Leeds & Liverpool Canal. Food Wed-Sun from noon. BB1 5BY
THIRA - Darwen Street. Tel: 01254 485641. Eat in take-away Indian open daily from noon. BB2 2BY
TCK DELI - Blackburn Market. Tel: 01254 669142. Vivacious market outlet offering an array of 'exquisite halal cuisine' to either eat in or take-away. BB1 5AF

Shopping
Tesco (Bridge 104A) and Asda (Bridge 101) have supermarkets near the canal, but the *tour de force* is Blackburn Market open daily (ex Sun) between 9am and 5.30pm. Occupying the ground floor of the Mall shopping centre, all retail life can be found within its high, daylight-bathed walls, not least a tripe stall.

Things to Do
CATHEDRAL - Cathedral Close. Tel: 01254 277430. Mother church in a diocese which covers most of Lancashire. The nave dates from 1826, though cathedral status was not granted until a hundred years later. Particularly distinctive (when viewed from the canal) is the lantern tower of 1967. BB1 5AA
MUSEUM & ART GALLERY - Museum Street. Tel: 01254 667130. Open Wed-Sat noon to 4.45pm. Industrial history and Victorian art. BB1 7AJ

Connections
BUSES - service 2 provides a useful hourly link for towpath walkers with Chorley, not directly connected by rail. Tel: 0871 200 2233.
TRAINS - Northern services to/from Burnley, Leeds, Manchester, Preston etc. Tel: 0345 748 4950.
TAXIS - 24/7. Tel: 01254 694949.

Cherry Tree
Map 28

Once the western terminus of Blackburn's trams, and characterised by bay-windowed terraces strung out along Preston Old Road, Cherry Tree offers alternative 'out of town' overnight moorings to Blackburn. Vanished beneath new housing on the offside west of Bridge 95 stood the premises of the Cherry Tree Machine Co. who specialised in the manufacture of laundry equipment.

Shopping
Sainsbury's Local and launderette.

Connections
TRAINS - hourly Northern service (bi-hourly Sun) to/from Blackburn and Preston. Tel: 0345 748 4950.

Feniscowles
Map 29

Western extremity of Blackburn's urban sprawl.

Eating & Drinking
FOUNDRY TAP - Livesey Branch Road (east of Bridge 93B). Tel: 01254 203200. Formerly the Oyster & Otter, this modern and stylish bar/restaurant serves food from noon daily throughout. BB2 5DQ

Shopping
Post office stores on A6061.

Connections
BUSES - see Blackburn service 2.

Riley Green
Map 29

Popular overnight mooring spot within easy reach (ideally by bicycle) of Hoghton Tower, a 12th century manor house immortalised by King James I's 'knighting' of a piece of beef there in 1617, thus giving rise to the term 'sirloin'. Tel: 01254 852986 for opening times and further information.

Eating & Drinking
HOGHTON ARMS - Withnell Road. Tel: 01254 201083. Family friendly Marston's pub about two-thirds of a mile south-east along the pavemented A675. Open from noon daily. PR6 8BL

CROSSING the watershed between the rivers Lostock and Darwen - both tributaries of the Ribble, though not destined to meet before the estuary - the canal traverses a largely rural landscape whose only brush with urbanisation is the suburb of Feniscowles. Little evidence remains of the Lancashire Union Railway which shadows the canal all the way to Wigan. At Feniscowles station there were sidings for a jam factory. Deep in woodland, an aqueduct carries the canal across the River Roddlesworth, a tributary of the Darwen.

As publishers, papermaking is a subject close to our hearts, though it is hardly any longer a significant facet of British industry. Many mills have closed. Those that haven't tend to be in foreign ownership. East Lancs was a notable centre of papermaking and the Leeds & Liverpool played an important role in transporting raw materials and fuel to the works and the finished goods away. Mooring rings by Bridge 93 are where coal came in by short boat

to feed the hungry furnaces of the Sun and Star paper mills. There is also evidence of an aerial ropeway employed to convey the coal down from the canal. But the mills have gone and their site has been taken over by a generating plant operated by Scottish Power of all people. Lucky they voted to stay in the UK! Old maps suggest that Blackburn's Contagious Diseases Hospital stood alongside Bridge 91B, though the remaining buildings have the look of warehousing. Furthermore, would they really build a contagious diseases hospital slap bang beside a canal? Sometimes old maps throw up more mysteries than they solve. A hire base occupies the site of Crook's boatbuilding yard whose sleek short boats were well-liked by coal carriers, if not quite held in such esteem by firms whose craft regularly had to mix with larger vessels in dockland.

Opened in 1997, the M65 motorway cuts a swathe across the countryside, but

continued overleaf:

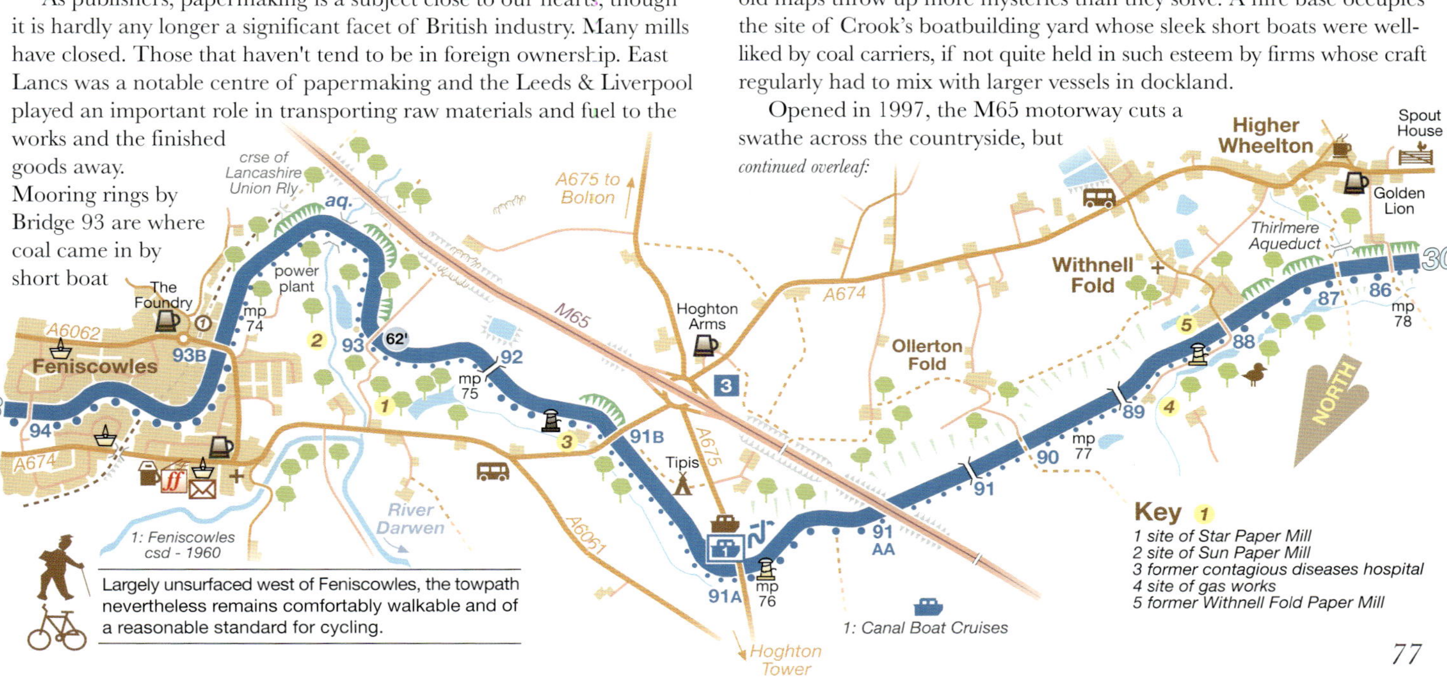

Largely unsurfaced west of Feniscowles, the towpath nevertheless remains comfortably walkable and of a reasonable standard for cycling.

continued from page 77:

does little to diminish the landscape's inherent tranquillity; why, even curlews were calling across the fields when we last passed. There are pleasant visitor moorings west of Bridge 91A in the lee of high ground. Curiosity aroused by the discovery of mooring rings alongside the house at Bridge 89, reference to the 6 inch Ordnance Survey map of 1909 revealed that it had been the site of Withnell Urban District Council's gas works. Part of the undying appeal of canal travel is the key it consistently offers you to unlock the past.

At Withnell Fold forty-eight hour visitor moorings exist in the shadow of a former paper mill, and

you could well fill those hours exploring the vicinity of this charming backwater. Beyond the towpath wall the land falls suddenly away to a series of filter beds which have been transformed into a nature reserve, a primeval paradise of dragonflies, water lilies and dipping pools. In the opposite direction a stroll up the cobbled lane from Bridge 88 introduces you to a 'model' village of paper-workers' cottages and a piquant little green with a sundial as its centre-piece; a moving memorial to the local men who didn't march home from two world wars, including one who was awarded the Victoria Cross.

Occupied by small businesses now, the paper mill was opened in 1844 by Thomas Blinkhorn Parke. Initially it produced tissue paper and newsprint, but after being absorbed by Wiggins Teape in 1890 the emphasis moved to high quality writing paper: Vandyke Border and Conqueror were produced here down the years. Coal to fire the machinery came in by short-boat. A model garden village community grew up around the mill to house its close-knit, Methodist workforce. Each cottage had a front garden, a backyard, wash-house, coal-house and privy. In the early years children began work at 8 years old and worked an 82-hour week. Women ceased work on marriage as their place was deemed to be at home. Entertainment came in the form of a Reading Room with a billiard room and an upper storey concert room with a sprung floor for dancing. In the 1930s a young pianist came regularly over from Blackburn to provide accompaniment at concerts and dances. Her name was Kathleen Ferrier, and she went on to become a world-famous contralto, though not before marrying a Withnell Fold man called Bert Wilson. The mill ran a cricket team (the village still does) and in 1932 the great West Indian all-rounder Learie Constantine (playing for Nelson in the Lancashire League at the time) appeared in a benefit match here. Its machinery outmoded, the mill ceased production in 1967.

By Bridge 87 a steep sided clough all but hides an enigmatic, high-vaulted bridge which might almost be carrying a second canal. No other guide, past or present, we could lay our hands on, even mentions it. Trust Pearsons to go, quite literally, the 'extra mile': it is the Thirlmere water supply aqueduct, opened in 1894 to bring water a hundred miles from the Lake District to Manchester. For the most part its course is subterranean, but another above ground section of it can be seen in Withnell Fold itself. The seven mile pound between Blackburn and Johnson's Hillock winds boskily - nay, heavenly - on.

POCKETS of sublime beauty mock the notion that the Leeds & Liverpool is only good-looking in its passage through the Yorkshire Dales. A canal of solitary disposition, gorse lines its banks, as it curves around the steep-sided valley of the River Lostock. Equally lonely herons patrol their beats. Rolling sheep pastures and ribbons of woodland combine to create a soothing landscape for the canal to lose itself in. Footpaths delve into woodlands beside the river and peace reigns supreme in what amounts to an unheralded vacuum of agricultural land between the old textile centres of East Lancs. Bridge 84 sports a throwback 'blue & yellow' number plate on its north face.

Wheelton Boat Club's linear moorings presage Johnson's Hillock Locks, which we've always considered one of the prettiest flights in the country. We had a yellow wagtail for company, and there wasn't a volunteer lockie in sight. An enchanting path runs along the offside: diminutive timber side-bridges spanning by-washes. Incidentally, these seven locks were built by the Lancaster Canal Co. whose main line comes in at the foot of the flight from its old summit at Walton, a good deal of its course having been submerged beneath the M61. At Walton cargoes were transhipped from barges onto tramway wagons and taken across the steeply-sided Ribble Valley before being arduously floated once again at Preston Basin. Poor old Lancaster, if it hadn't been so disjointed, and if it hadn't been callously severed in the Sixties by the M6 north of Carnforth, it might have been one of the canal system's *pieces de resistance*.

Milepost 80 appears to be an appurtenance of a tree, recalling the 'living mileposts' of the Old Union Canals in Leicestershire. North of Bridge 80 stood Whittle Springs Brewery on a site now occupied by prestigious housing. In the 19th century there were pleasure gardens here, a popular venue for Sunday School outings by suitably spruced-up barge. The loftily-spired Preston Temple of the Church of Latter Day Saints is a prominent landmark to the west. The post-textile fate of mills tends to lie in the lap of the gods. Canal Mill, prominent to the south of Bridge 79 had been the centre-piece of a

continued overleaf:

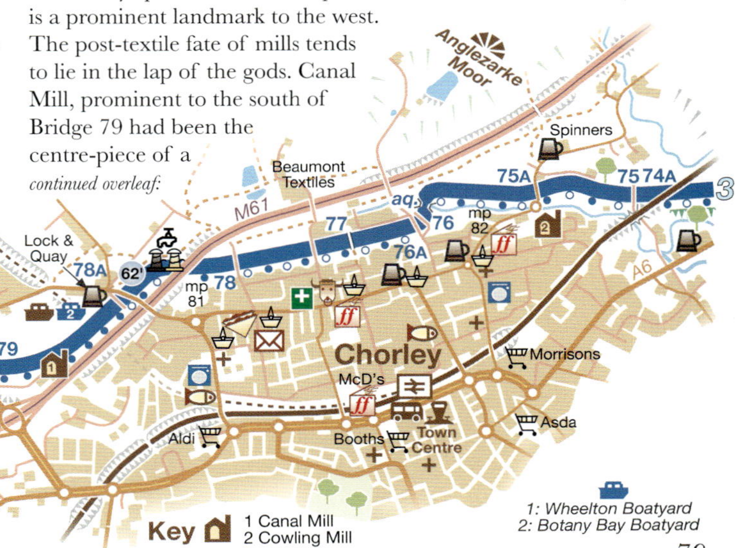

continued from page 79:

retail outlet called Botany Bay, a name said to have been given to the north-eastern outskirts of Chorley by disenchanted navvies. The centre closed in 2019 and redevelopment schemes have stalled. Meanwhile, the mill's future remains in the balance. Let's hope it doesn't go the way of the equally impressive, terracotta adorned Cowling Mill on the other side of Chorley, apparently in the process of being demolished.

A sensible (but hard won) arrangement between the Leeds & Liverpool and Lancaster canal companies saw the former employ the latter's route between Wigan and Heapey to avoid duplication. Not that the deal did the Lancaster much good, for it progressed neither southwards nor northwards, and pretty soon everyone forgot that it had built this length of the Leeds & Liverpool at all.

If the canal appears to cold shoulder Chorley, it's probably, as the L&L historian Mike Clarke explains on a number of occasions, that even in the 18th century land prices were higher in the centre of towns and canal promoters had no wish to incur unnecessary outlay. Plus there were contours to consider. Moreover, in this instance, it was the Lancaster Canal Company who built this stretch of canal, and their focus was very much on reaching Manchester with the help of the Bridgewater Canal beyond Worsley.

Followers of the oval ball game will be intrigued to know that Bill Beaumont's family textile firm (of which he was formerly CEO) go about their business in modern premises on the far side of the M61.

Wheelton   Map 30

Quaint village of redbrick terraces with an imposing clock-towered war memorial.

Eating & Drinking

MALTHOUSE FARM - Blackburn Road, Wheelton (adjacent Bridge 80). Tel: 01257 232889. Chef & Brewer/Premier Inn. Food served from 11.30am weekdays and from 9.30am weekends. PR6 8AB
WHINS GREEN KITCHEN - Whins Lane (access via Bridge 83). Tel: 01254 958311. Delightful cafe/restaurant, with a commanding view over the countryside, serving vegan, vegetarian and plant based cuisine. Open 10am-5pm daily ex Mon. PR6 8HN
Tea room in the village. The Dressers Arms (Tel: 01254 830041) on Brier's Brow is also well liked.

Connections

BUSES - Blackburn Bus Co. service 2 runs hourly between Chorley and Blackburn. Tel: 0871 200 2233.

Chorley   Map 30

A companionable town on the A6 noted for a dried fruit confection related to Eccles Cakes. Noted too, for its eponymous Pals, a company of the East Lancashire Regiment who, kitted out in blue uniforms woven locally, volunteered en masse for service in the First World War, suffering greatly for their patriotism on the Somme in 1916. St Mary's RC church was designed by Joseph Hansom of cab fame.

Eating & Drinking

LOCK & QUAY - Botany Bay (adjacent Bridge 78A). Tel: 01257 261816. Canalside pub refurbished 2019. Opening times vary. PR6 9AE
THE SPINNERS - Cowling Road (East of Bridge 75A). Tel: 01257 241622. Country inn on the flank of Anglezarke Moor. Food served Wed-Sun. PR6 9EA
Fish & chip fans will relish Pawsons on High Street.

Shopping

The town centre lies the best part of quarter of an hour's walk to the west. Aldi, Asda, Morrisons, and those paragons, Booths supermarkets, but more significantly, an excellent market which climaxes on Tuesdays and closes on Wednesdays to recover.

Connections

BUSES - see Wheelton.
TRAINS - frequent (electrified) Northern services to/from Preston and Manchester. Tel: 0345 748 4950.
TAXIS - Star Cars. Tel: 01257 279279.

Adlington   Map 31

An erstwhile textile and coal mining community, whose main employer now produces machinery for renewable bio-power plants. The village is also the most practical canalside base camp for the ascent of Rivington Pike a couple of country miles to the east. Somewhat surprisingly, the village has a railway station.

Eating & Drinking

BISTECCA - Market Street. Tel: 01257 474666. Italian restaurant open from 5-9pm Tue-Fri, 4-9pm Sat and 3-8pm Sun. PR7 4HE
JR's ALE HOUSE - Church Street. Tel: 0758 388 5560. Micropub serving up to four real ales. PR7 4EX
THE RETREAT - Church Street. Tel: 01257 481894. Bar & grill housed in old church. Food served from noon daily. PR7 4EX
Plus pubs, take-aways, and a cafe at White Bear Marina.

Shopping

Premier store at T-junction 200 yards from Bridge 69. Also nearby are a pharmacy and post office. Don't forget to call in at Frederick's (Tel: 01257 263154) Italian ice cream dairy easily accessible on foot from Bridge 73 and open 9am-10pm daily. PR7 4AL

31 LEEDS & LIVERPOOL CANAL Adlington 4mls/0lks/1½hrs

LOCK-free, and crossing the boundary between Greater Manchester and Lancashire - a spurious frontier unrecognised before the intervention of Edward Heath and his pettifogging ministers in 1974 - the 'Lancaster Pool' essays a serpentine course: admit it, you never thought the canal would be this beguiling between Chorley and Wigan.

At Bridge 73 boaters become the cynosure of ice-cream-cone-licking spectators, customers of a firm of Italian descent called Frederick's who've been pandering to sweet-toothed Chorleyians since 1892. It's hard to imagine in our motorway-dominated age that the A6 was once the best way of getting from London to Carlisle by car. An impressive girder bridge carries it across the canal at Heath Charnock. The course of an abandoned railway leads to the site of Duxbury Park and Ellerbeck collieries. The former was closed by its owners during the Depression before being rescued and re-opened by its workforce, thereafter remaining open until 1965. So much coal was carried along the canal that, when the water was drained for maintenance, local people would rush out and pick (or 'kebb' in the local vernacular) any spilt coal from the bed of the canal.

An abiding image of Adlington is of a factory chimney smoking over slate rooftops with a fine swell of Lowryesque moorland to the rear. Wart-like Rivington Pike is backed by Winter Hill with its radio masts. Pike Tower has been an East Lancashire landmark since it was erected in the 18th century. It stands on a summit 1200ft above sea level and on a clear day you can see the Isle of Man from here.

The River Douglas (which eventually becomes navigable - see Maps 34A/B) is bridged by an aqueduct; after heavy rain the river water gushes down between its high banks like the Charge of the Light Brigade. Arley Hall, nowadays the club house for Wigan Golf Club, is surrounded by a moat. Another aqueduct conveys the canal across the trackbed of the Lancashire Union Railway (or NCR 55 as it has commendably become).

Pleasant visitor moorings are provided to the north of Bridge 63 where there are remains of an old coal tippler, though the accompanying pub, disconsolately boarded-up for years, appears beyond redemption.

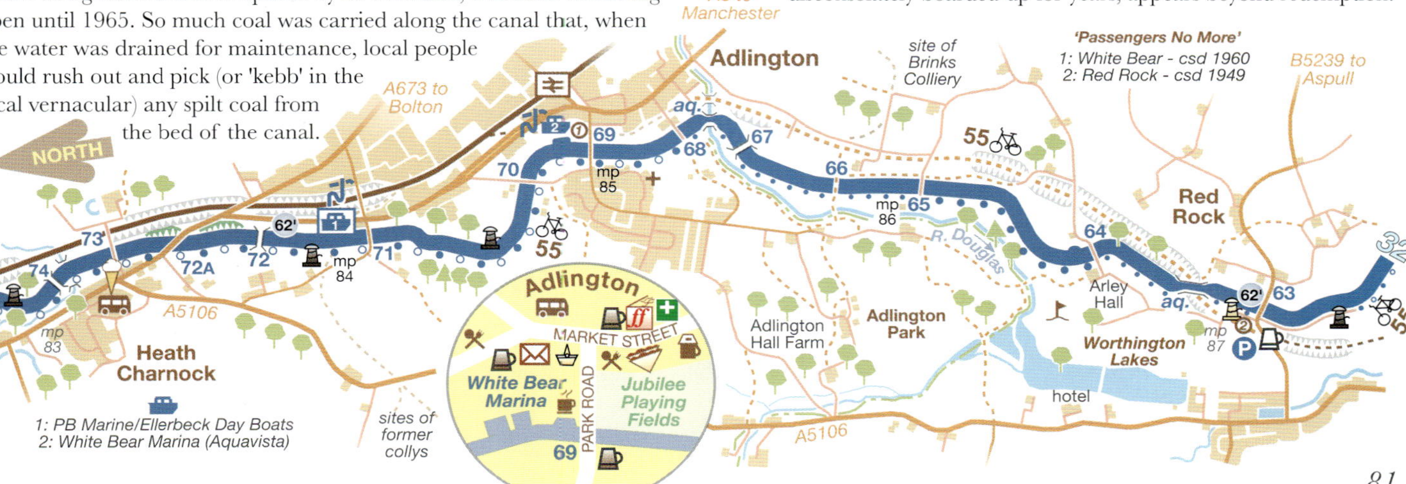

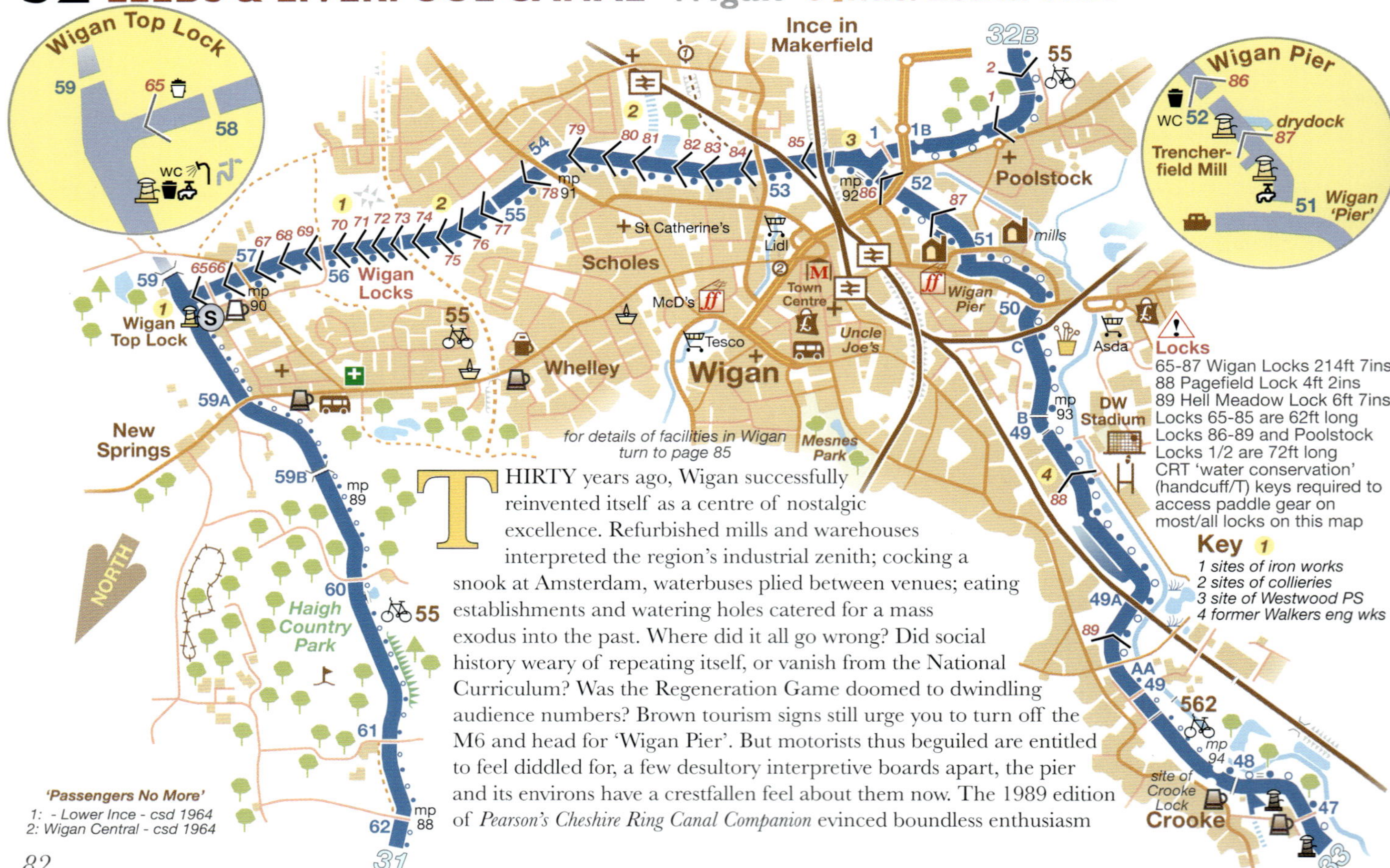

for details of facilities in Wigan turn to page 85

THIRTY years ago, Wigan successfully reinvented itself as a centre of nostalgic excellence. Refurbished mills and warehouses interpreted the region's industrial zenith; cocking a snook at Amsterdam, waterbuses plied between venues; eating establishments and watering holes catered for a mass exodus into the past. Where did it all go wrong? Did social history weary of repeating itself, or vanish from the National Curriculum? Was the Regeneration Game doomed to dwindling audience numbers? Brown tourism signs still urge you to turn off the M6 and head for 'Wigan Pier'. But motorists thus beguiled are entitled to feel diddled for, a few desultory interpretive boards apart, the pier and its environs have a crestfallen feel about them now. The 1989 edition of *Pearson's Cheshire Ring Canal Companion* evinced boundless enthusiasm for Wigan Metropolitan Borough Council's achievements along the town's canal corridor. It gives us no satisfaction to be unable to echo those sentiments now.

And yet, for all the retrenchment, Wigan remains a stimulating place to visit - especially by canal - and the Leeds & Liverpool's (alright, Lancaster's if you want to be pedantic) approach past Haigh (pronounced 'Hay') Country Park keeps urbanisation at bay until the last possible moment. Water lilies thrive in the vicinity of Bridge 61. The main Euston-Glasgow railway is momentarily visible in the valley of the Douglas, and the spire of Standish church punctures the horizon. The canal rides along a low escarpment providing views westwards as far as the redundant cooling towers of Fiddler's Ferry power station on the banks of the Mersey. The rhododendrony grounds of Haigh Hall spill down to the waterside. There is a serene passage through woodland planted, apparently, by unemployed millworkers during the American cotton famine. An elegant footbridge throws its graceful, lattice span over the cut adjacent to a small, reedy basin replete with stone side bridge, probably provided for goods destined for use at the hall. Haigh Hall itself is now a municipal amenity, more or less masked by trees beyond a golf course which accompanies the canal for some way. By Bridge 59A there are suggestions of a former wharf and, nearby, a curiously bell-towered stone house.

Wigan Twenty-One

On hearing the words "Wigan Twenty-One ...", you half expect Eddie Waring to complete the statement ... "Featherstone Rovers Nil". But the twenty-one are not a rugby league scoreline, they are a reputedly daunting flight of locks which carry the Leeds & Liverpool Canal down around the south-eastern periphery of Wigan, descending two hundred feet in the process. Sixty-two feet in length (as opposed to the 72ft of the locks west of Wigan) they are the sort of flight that boaters have nightmares about. But, as so often with bad dreams, you awake to a relieved reality. As ever, the more crew you have at your disposal, the less gruelling a flight will be. Moreover, it goes without saying, that narrow boats sharing the chambers,

waste less water and halve the workload. Handcuffed (like criminals) the locks are at their densest in the middle of the flight, or perhaps it is just that morale is at its lowest then. Where space is restricted, one or two of the locks have winches to operate the balance beams. Gradually you begin to derive a perverse sense of enjoyment from the experience - the stirring stuff of tales like: "We did Wigan in under four hours!"

Top Lock is also known as Aspull Lock. It is overlooked by a lock-keeper's cottage and, to the rear, old stabling. Spend a while exploring. Go and gaze over the parapet of Bridge 59, along what was to have been the line of the Lancaster Canal on its way to join the Bridgewater at West-houghton near Worsley, before the money ran out.

But sooner or later you'll have to take the plunge, entering Lock 65 and commencing the long descent. Bridge 56 is inscribed '1816' and gives access to Kirkless Hall, a partially half-timbered farmhouse predating the industrialisation of Wigan. Some of the lock walls bear Roman numerals suggesting that the flight was originally numbered from the top as opposed to Leeds as is the case now.

A grassy slag heap looms over Lock 73. The locals call it 'Rabbit Rocks' and it was made from the waste material of Kirkless Iron & Steel Works which occupied the neighbouring wastelands between 1858 and 1931. Scramble to the top for panoramic views over Wigan and northwards to Winter Hill and Rivington Pike.

The slender, slightly kinked spire of St Catherine's, Scholes proves a stubborn landmark, seemingly in no hurry to come closer or grow distant depending on your direction of travel. St Patrick's amateur rugby league ground, and little streets with lingering whiffs of the 19th century about them, overlook the canal between locks 80 and 81. Long ago an arm led to Moss Hall and Ince Hall collieries from the tail of Lock 81.

Four railways spanned the flight, though only two remain. Adopted as part of National Cycle Route 55 - which links Ironbridge in Shropshire with Preston - the Lancashire Union crossed the canal between locks 73 and 74. Once upon a time banana trains from Garston Dock steamed

continued overleaf:

continued from page 83:

across the canal here, heading for Scotland with a vital dietary antidote to Tunnock's Caramel Wafers. The Great Central bridged the canal between locks 83 and 84. Lock 85 is framed by two railways: the old Lancashire & Yorkshire's route to Manchester Victoria (53B) and the (ex LNWR) West Coast Main Line (53D). Curving round to its junction with the Leigh Branch, the main line passes the redeveloped site of Westwood Power Station which received coal from mines in the vicinity of Leigh until 1972.

The Leigh Branch

Maps 32A/B depict the route of the seven mile Leigh Branch, an important link for north-south boaters, and increasingly popular with the opening of the Liverpool Link. As built, circa 1820, Poolstock Locks were 62 feet long, precluding their use by narrowboats trading up from the Manchester area. Responding to criticism from Pickfords and other carriers, they were extended by ten feet a couple of years later. The branch is a worthwhile detour for travellers on the main line, an atmospheric journey through a landscape reclaimed by nature from the ravages of coal mining.

Main Line to/from Liverpool

The extra length of the locks west of Wigan is immediately apparent to anyone who has come down the Twenty-one. The Canal & River Trust's regional office at the back of Trencherfield Mill was 'temporarily closed' when we went to press. Between locks 86 and 87 a drydock is all that remains of a once extensive boatbuilding yard. The last wooden built L&L boat, *Darlington*, was launched here in 1951. Trencherfield Mill's steam engine performs on selected dates. Handsomely built from

Scalped bystander, Wigan

stone and boasting twin arched loading bays, the original terminal warehouse of 1777 can be seen on your right as the canal dog-legs beneath Bridge 51, where the sculpture of a cloth-capped onlooker appears to have been 'scalped' - symbolically or otherwise, we couldn't decide. At this point, the canal enters the area regenerated in the 1980s but abandoned, since 2007, to its own, increasingly threadbare devices; though there was evidence, on our most recent research trip, of the warehouses being fitted out as 'eco townhouses'.

Generally ascribed to music hall artiste, George Formby senior, the visual counterpart of the Wigan Pier joke is thought to have been either an extensive tramway trestle bridge spanning the Douglas Valley, or a landstage where coal wagons had their contents tipped into barges. The reconstructed loading 'pier' opposite the abandoned visitor centre is conveniently regarded as Wigan Pier now, with an interpretive board on hand to offer moral support. At least having reached this far you can claim to have stumbled upon it, unlike George Orwell who allegorically failed to do so when writing his 1938 polemic *The Road to Wigan Pier*.

Passing beneath the Lancashire & Yorkshire Railway's once prestigious Manchester Victoria to Liverpool Exchange line (trains run only as far west as Kirkby now where they connect with Merseyrail), the canal slowly begins to disentangle itself from Wigan's clutches. The towpath side of the canal is dominated by the 25,000 seater DW Stadium, home to Wigan Athletic Football Club and Wigan Warriors Rugby League Club. DW stands for Dave Whelan, former professional footballer and self-made businessman who famously broke his leg playing for Blackburn Rovers in the 1960 FA Cup final against Wolverhampton Wanderers. In fitting, if belated, consolation, 'The Latics' themselves won the cup in 2013.

Pagefield Lock dates from 1913. Alongside it stood a large ironworks which once belonged to Walkers, manufacturers - amongst other more mundane engineering items - of dainty diesel railcars for the West Clare and County Donegal narrow-gauge railways in Ireland. Crooke Lock was abandoned following subsidence, but it is still necessary to pass through the narrows formed of its chamber. Coal was still being loaded for Liverpool's gas works and Wigan's Westwood power station from a bankside chute (offside, just west of Bridge 47) onto boats until 1964. Crooke was a mining village personified. Primitive steam locomotives were working along tramways in the vicinity as early as 1812. A branch canal, used for moorings now, previously linked the main line to an 1,100 yard tunnel which took boats right to the coal face.

Wigan   Map 32

Is there a perverse equation between sheer downright ugliness and native warmth and wit? If Wigan is anything to go by, there is, and we always look forward to revisiting this idiosyncratic town with every update. In common with most industrial towns, the handsomest buildings are Victorian and Edwardian. With time at your disposal, make it your business to visit the Mesnes Park with its Coalbrookdale-cast fountain, bandstand, Boer War memorial and lavish pavilion from which Frederick's (of Heath Charnock, Map 31) dispense 'decadent' ice cream sundaes.

Eating & Drinking

FRANCO'S - Rodney Street. Tel: 01942 248668. Nicely furnished Italian open daily from 5.30pm (4pm Sun). WN1 1DG

GALLIMORE'S - The Wiend. Tel: 01942 492100. Fine-dining daily from 11.30am ex Mon evenings. WN1 1PF

RED DOOR BISTRO - College Avenue (off Library Street). Tel: 01942 820484. Likeable bistro open Tue-Sat lunchtimes and Thur-Sat evenings. WN1 1DF

SWAN & RAILWAY - Wallgate. Tel: 01942 375817. Classic Victorian town pub: opens noon. WN1 1BA

WIGAN CENTRAL - Queen Street. Tel: 01942 246425. Marvellous *Good Beer Guide* listed, railway themed real ale & cider bar beneath the arches of North Western station and easily reached from the canal. Open from noon daily (11am Sat). WN3 4DY

Shopping

Asda, Lidl and Tesco supermarkets, though none particularly close to the canal. Grand Arcade and Galleries shopping centres. The excellent market hall is open daily from 8.30am ex Sun. The Makinson Arcade of 1898 is as memorable for its soaring ironwork as its shops. You will not want to leave town without a tin of Uncle Joe's Mint Balls from their factory shop on Dorning Street - Tel: 01942 243464.

Things to Do

MUSEUM OF WIGAN LIFE - Library Street. Tel: 01942 828128. Displays of Wigan history from Roman times through its industrial zenith to the phenomenon that was 'Northern Soul', with a bit of Rugby League en route. Open Tue-Sat 10am-4pm. Shop. WN1 1NU

TRENCHERFIELD MILL - Heritage Way. Tel: 01942 828128. Restored mill engine open to the public on selected dates. WN3 4EF

Connections

BUSES - Arriva service 362 links Wigan with Chorley; Stagecoach services 8/9 with Leigh. Tel: 0871 200 2233. TRAINS - inter-city services north and south (London Euston, Birmingham etc) from North Western station; useful local services along the Douglas Valley calling at Appley Bridge, Parbold, Burscough Bridge etc (and thus ideal for one-way towpath walks) from Wallgate station. Tel: 0345 748 4950.

TAXIS - Blue Star. Tel: 01942 242424.

Leigh   Map 32A

A surprisingly substantial town, which has a longer pedigree than its prevailing image of a mill and mine community implies. The noble town hall dates from 1907 and is, according to the local publicity handout, "a dignified expression of civic prosperity and pride," and we wouldn't quibble. Elsewhere, copious use of red Accrington brick, terracotta and Portland stone evince an Edwardian determination to compete with Wigan, Bolton and St Helens. The town centre and all its facilities lies just north of Leigh Bridge, by which there are ample moorings. James Hilton, author of *Goodbye Mr Chips*, *Lost Horizon* and *Random Harvest*, all made into Hollywood films, was born here in 1900.

Eating & Drinking

DON ALBERTO'S - Loom Retail Park. Tel: 01942 673858. Homely Italian open from 10am. WN7 4BA

Shopping

Aldi, Tesco and Lidl supermarkets either side of Leigh Bridge. A fine indoor market adjoins the Spinning Gate shopping centre. Taylors pies are a local delicacy.

Connections

BUSES - Stagecoach services 8/9 link Leigh with Wigan, a useful facility for towpath walkers. In the absence of a railway station Firstbus V1 operates at frequent intervals to/from Manchester via the Leigh-Ellenbrook guided-busway. Tel: 0871 200 2233.

IKE spies meeting on a park bench, the Bridgewater Canal and Leigh Branch of the Leeds & Liverpool Canal surreptitiously exchange indentities. Once this most unobtrusive of junctions was overlooked by a substantial canal warehouse and a ropeworks. The warehouse has become a (now closed) pub, the ropeworks a housing zone: the glory, as the Arabs so eloquently put it, has departed. One minute the bridges have names, the next numbers. Otherwise nothing seems to ruffle the stolid waters of the two canals. Pearson's *Cheshire Ring Canal Companion* provides coverage of the Bridgewater Canal's progress southwards to Manchester and beyond.

No need to put your name speculatively down for interplanetary travel, you can experience authentic Martian landscapes merely by exploring the Leigh Branch. Opened in 1820 to connect the Leeds & Liverpool and Bridgewater canals, packet boats briefly offered the most expedient means of travelling from Manchester to Liverpool until Stephenson's new fangled railway rendered them obsolete ten years later. For a hundred and fifty years, however, the Leigh Branch was fully employed in carrying coal. In 1972 the architectural critic and broadcaster, Ian Nairn, made a half hour programme for the BBC called *Trans-Pennine Canal*. His cameraman filmed a sequence following a coal-laden barge on the last lap of its six

mile shuttle from Plank Lane to Wigan's Westwood Power Station, an everyday scene that the usually perceptive Nairn omitted to comment on, probably because the soundtrack at that point was given over to some crass folksong. Before the film was broadcast the traffic had ceased. Allegedly because the coal from Bickershaw Colliery was no longer burning well in Westwood's furnaces. But that was the sort of drivel nationalised industries used to propagate when it suited them to stop doing something.

In its commercial hey-day, coal for Liverpool was worked in both directions along the Leigh Branch: up to Wigan and over to Liverpool; or down along the Bridgewater to Runcorn and thence via the Mersey. For the latter traffic, Mersey flats were employed; horse-drawn on the canals and sailed across the estuary.

Nowadays, the only hints that this was once a coal mining district are the embanked canal and the lakes (or 'flashes') which accompany it. As subsidence undermined the landscape it was necessary to keep raising the

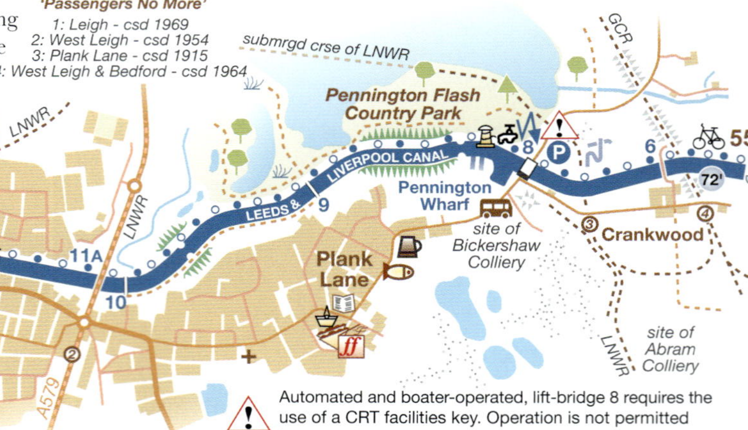

Automated and boater-operated, lift-bridge 8 requires the use of a CRT facilities key. Operation is not permitted between 8-9.30am and 4.30-6pm Mon-Fri.

canal banks, usually with pit waste, and as a result the canal is deeper than usual. Similarly, as the land fell away, pockets of water gathered which have been absorbed into the landscape as if they were natural features. On sunny days Pennington Flash glitters with sailing dinghies. 'Mars', as the posters say, 'can be surprisingly pretty'!

For a town of forty thousand souls, it seems bizarre that Leigh no longer has a railway station. Manchester, twelve miles to the east, is the best part of an hour away by the fastest ('guided') bus. And the politicians blithely cant about 'levelling-up'. Never mind the numerous mineral lines, four distinct railways crossed the canal depicted on the accompanying map, the last one losing its passenger trains in 1969. What an era of endings the late Sixties and early Seventies represented. In the pre-grouping period three of the lines belonged to the London & North Western Railway, the dominant force in the area, but the fourth was a daring raid into rival territory by the Great Central, and the overgrown earthworks of that line can be traced at Crankwood. The route of the earliest public railway in the district - the Bolton & Leigh, opened for goods in 1828 and three years later to passengers - lies under the A579 which crosses the canal on Bridge 11A. As can be imagined, there was a good deal of competition for traffic between canal and railway, not all of it above board. When this length of canal was first featured in the Canal Companions, Bickershaw Colliery was still in business, it closed in 1992. Coal trains squealed down rickety tracks to collect its output,

Bridge 6, Crankwood

and we managed to juxtaposition a photograph of our research boat - steered by a windswept individual in a yellow sou'wester - against a backdrop of a trio of gaunt headstocks and their spinning wheels. Vanished! And in their place, a development scheme of 650 new houses and a marina.

continued from page 88:
earlier electric tramway.

Overhead wires do, however, hang above the railway tracks of the West Coast Main Line where it bridges the canal adjacent to the site of Bamfurlong Colliery, now just a wasteground of scrub. A succession of rusted wagon tipplers - decaying more and more to the point of oblivion each time we pass this way - point to where coal was loaded onto boats in days gone by. Between Bamfurlong and Poolstock (Map 32) no roads, nor any houses, come remotely near the canal.

On the skyline ahead, Wigan rears up like a post-industrial Lindisfarne and you feel you are travelling towards it across the quicksands of history. A sort of inverted archipelago of subsidence induced lakes, or flashes, lines this 'Water Road to Wigan Pier', bearing contradictory comparison with the Shropshire meres that border the altogether more picturesque Llangollen Canal. Extensive reed beds and nesting colonies delight the birdwatcher latent in us all. Pearson's Flash derives its name from the Pearson & Knowles Coal & Iron Co who had their finger in a lot of pies locally in the early 20th century; much like gravy-fingered Lock-wheeler if the truth be known.

TONGUE in cheek, we draw spurious comparisons with Mars, but regular Pearson's users will know that this is precisely the kind of time-ravaged, world-weary topography that excites our muse the most. Everywhere you look, along this apparently bland length of canal, an exciting story cries out to be told, much of it relating to Old King Coal, but there are other nuances as well.

Subsidence - as you will have already seen travelling northwards - has wrought havoc with the landscape: cause and effect incarnate. Fluctuating levels necessitated removal of two locks at Dover, though the canal still narrows as it passes through the redundant chambers. The aqueduct which carries the canal over Hey Brook reveals alterations too. A cindery path leads up to Abram's Austin & Paley church, consecrated in 1937. At that time, Warrington Road (A573) was framed by mine-workers' terraces, but housing estates now occupy the sooty pastures which surrounded Maypole Colliery. The mine derived its name from the nearby Morris Dancers' Ground, the origins of which are shrouded in the mists of time, but it is

a tradition revived of late on the last Saturday in June.

Both the Maypole and Abram (Map 32A) pits were the scenes of underground explosions: 18th August 1908, 75 fatalities; 19th December 1881, 48 fatalities respectively. Mining - it hardly goes without saying - could be a heart-rending occupation. One is reminded of Philip Larkin's poem *The Explosion*: the oath-edged men in pitboots, shouldering off the freshened silence.

Bridge 3 carries the A58 across the canal, one of those intriguing pre-motorway-age numbers whose extremities are as obscure these days as mines and morris dancing: Prescot (St Helens) to Wetherby is the reality in this case. Between 1930 and 1958 this road carried the wires of the extensive South Lancashire Transport trolleybus network across the canal. Totalling more than 35 route miles, and linking such disparate industrial communities as Ashton-in-Makerfield, Hindley, Leigh, Swinton, Farnworth and Bolton, the company's fleet was notable for the longevity of its red-liveried vehicles, a number of its six-wheel, Roe bodied Guys remaining in use for the whole lifespan of the operation which had superseded an

continued on page 87:

S EWN into the exquisite embroidery of the Douglas Valley, the Leeds & Liverpool Canal lulls one into a fallacious vacuity; a faux conviction that this beguiling stretch of waterway cannot plausibly be the work of vulgar 18th century canal builders, but rather the landscaping of a Capability Brown or a Humphry Repton. It was May when we most recently had the privilege of traversing this valley: buzzards were wheeling over the watermeadows; flotillas of mallard ducklings zigzagged their way across the canal; and hawthorn bushes attended the towpath like bridesmaids attired in creamy lace. It would be difficult to find a lovelier length of inland waterway to explore. Yet all the way from Wigan to Burscough we only encountered four boats. The Leeds & Liverpool's comparative unpopularity is impossible to fathom.

Above Dean Lock the canal is crossed in quick succession by two subsequent eras of transport. The Wigan-Southport railway (still mechanically signalled at the time of writing) was completed in 1855; the Warrington-Preston section of the M6 a hundred and eight years later. Gathurst Viaduct took four years to build. Industry past and present borders the canal. On the north bank stood an extensive works which produced explosives for the mining industry, whilst visible to the south stands Heinz's baked bean factory: up to three million cans a day.

The canalscape at Dean Lock is compelling. As construction of the canal proceeded eastwards, a lock connected to the Douglas at this point, enabling traffic to reach Wigan and providing a source of water. You can trace its site by the tail of the present lock beside the end wall of the house which was once a toll office. Though the Douglas Navigation was abandoned following completion of the canal to Wigan in 1781, the lock remained in use thereafter to access wharves on the river served by tramways delivering coal from Orrell. It was finally abandoned towards

continued overleaf:

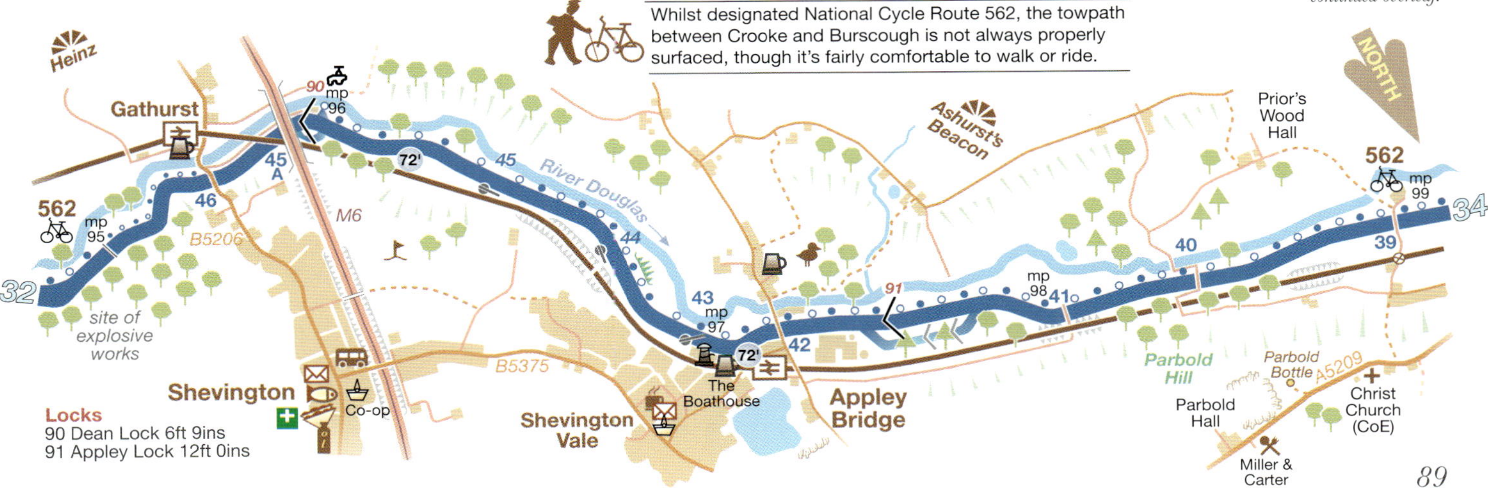

continued from page 89:

the end of the 19th century. A third lock was provided to duplicate the existing lock at the height of the canal's success, but whilst this retains its gates, it has been out of use for many years.

At Appley Lock you complete your descent from (or commence your climb towards) the Leeds & Liverpool's summit at Foulridge. Forty-seven notches on your tiller. Thank goodness they weren't all as intimidatingly deep as Appley's dozen feet. The adjoining duplicated pair, subsequently constructed to improve water supply, and expedite the hectic coal trade, are long abandoned. Swing-bridge 43 is manually operated but requires a CRT facilities (handcuff/T) key to unlock it. Swing-bridges 44 and 45 have been quietly disintegrating for a number of years.

The canal is overlooked to the south by Ashurst Hill (570ft), and to the north by Parbold Hill (400ft). Ashurst's Beacon peeps over the tree-line. In the 16th century there was a chain of beacons all the way from Liverpool to Lancaster. Ashurst's was rebuilt in 1798 by Lord Skelmersdale, and would be lit in the event of a Napoleonic invasion. Parbold Hill boasts its own 'folly' in the shape of a small port-bottle-shaped monument to the Reform Bill of 1832. On the second Saturday of February Skelmersdale Boundary Harriers (founded in 1898) organise a very muddy, obstacle-strewn race to the top and back. Parbold Hall is said to have been the original home of Ampleforth School before it decamped to the North Riding. On a clear day you can see Blackpool Tower.

Appley Bridge — Map 33

Eating & Drinking

THE BOATHOUSE - Mill Lane (adjacent Bridge 42). Tel: 01257 255505. Comfortably refurbished pub/kitchen open from noon Wed-Sun. Food served lunchtimes until 3pm and from 5pm for dinner. Sunday menu 12-7pm. WN6 9DA

Shopping

Post office stores at nearby Shevington Vale, and additional facilities at Shevington itself via B5375.

Connections

TRAINS - Northern services to/from Burscough Bridge and Southport, and Wigan Wallgate and Manchester. Tel: 0345 748 4950.

Parbold — Map 34

An apparently flourishing village, Parbold boasts two spired churches: Roman Catholic Our Lady - together with the adjoining Convent of Notre Dame - taking pride of place in the village whilst, unusually, the Anglican Christ Church stands cold-shouldered on the neighbouring hillside. Indeed, Catholicism seems to have radiated from Liverpool, an Irish diaspora perhaps? On a humbler scale look out for the delightfully-gabled Women's Institute, the deeply-eaved signal cabin, and an ornately-facaded bank which has become a dress shop: checks instead of cheques!

Eating & Drinking

MILLER & CARTER - Parbold Hill. Tel: 01257 462318. Steak bar and grill offering vistas over the Douglas Valley if you're lucky enough to get a window table. Open daily from 11.30am. WN8 7TG

WAYFARER - Alder Lane (A5209). Tel: 01257 464600. Charming *Good Beer Guide* listed brew pub with dining rooms. Their beer is delightfully known as Problem Child and excellent food is served lunch and dinner weekdays and from noon at weekends. WN8 7NL

THE WINDMILL - Mill Lane (adjacent to Bridge 37D). Tel: 01257 462935. Handsome whitewashed pub close to the canal. Food lunch and evening through the week and from noon at weekends. WN8 7NW

YOURS IS THE EARTH - Mill Lane (adjacent Bridge 37D). Tel: 01257 462999. Cafe/deli/ice cream parlour/gift shop open Thur-Mon 10am-5pm (4pm Sun). Breakfasts, soups, sandwiches etc; afternoon teas and cakes. ('YITE' to locals!). WN8 7NW

Also: coffee shop, Tandoori, Chinese/fish & chips take-aways.

Shopping

Good little shopping centre whose facilities include a Morrisons convenience store which incorporates a Post Office; deli, fruit & vegetable shop, pharmacy, DIY store and a butchers called Reynolds (ex Thur & Sun) who enterprisingly make their own pies.

Connections

BUSES - service 313 operates hourly (ex Sun) to from Skelmersdale. Tel: 0871 200 2233.

TRAINS - see Appley Bridge

Newburgh — Map 34

Pretty village with a conservation area at its heart.

Eating & Drinking

EDEN - Course Lane. Tel: 01257 581998. Tea room and gallery from 9am (10am Sun). WN8 7UB

RED LION - Ashbrow. Tel: 01257 462336. Well-appointed country pub owned by Marston's. Open from 11am daily. Accommodation. WN8 7NF

Tea rooms at the post office.

Shopping

Wilton House Farm Shop open daily 10am-5pm.

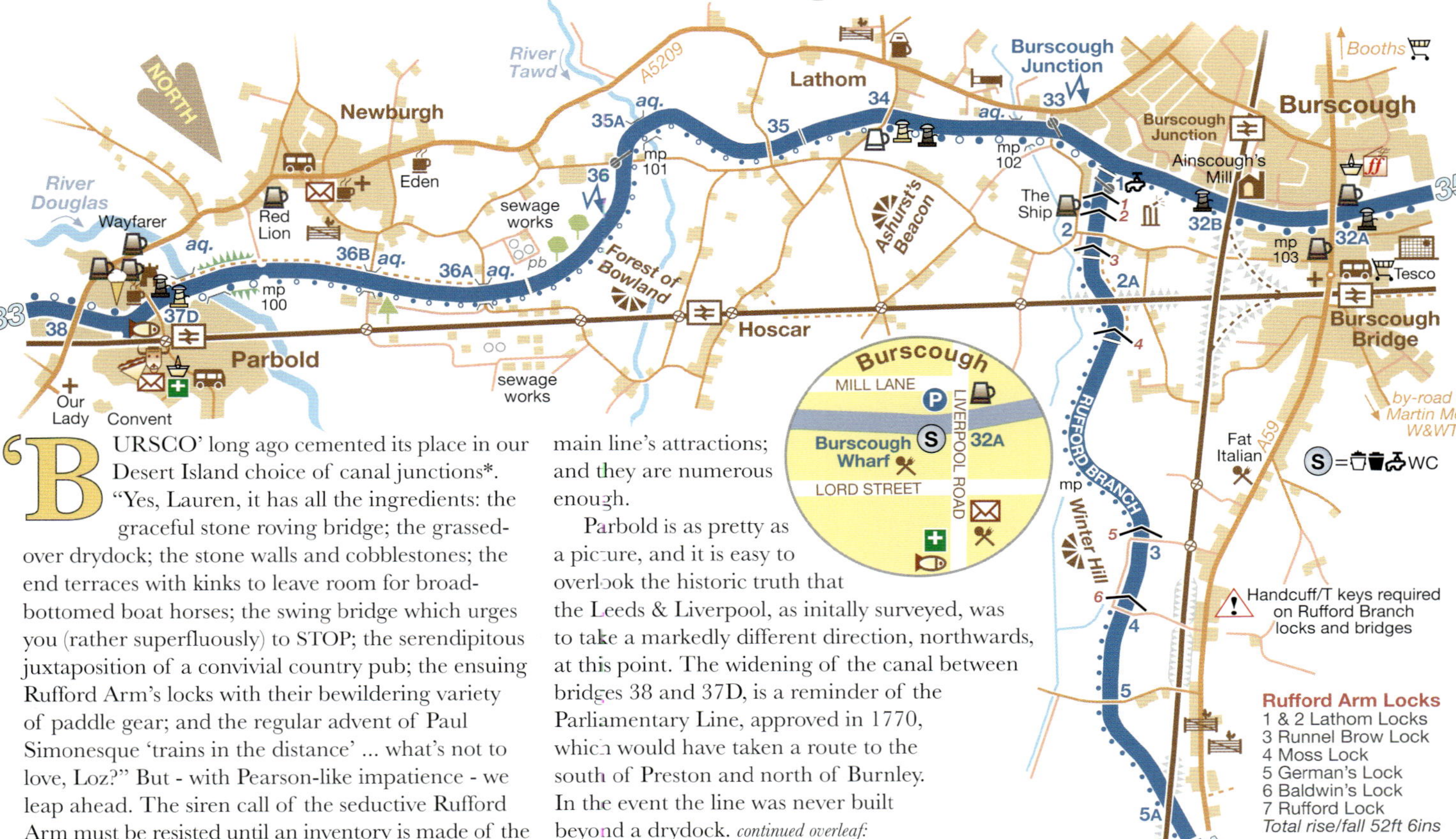

'B URSCO' long ago cemented its place in our Desert Island choice of canal junctions*. "Yes, Lauren, it has all the ingredients: the graceful stone roving bridge; the grassed-over drydock; the stone walls and cobblestones; the end terraces with kinks to leave room for broad-bottomed boat horses; the swing bridge which urges you (rather superfluously) to STOP; the serendipitous juxtaposition of a convivial country pub; the ensuing Rufford Arm's locks with their bewildering variety of paddle gear; and the regular advent of Paul Simonesque 'trains in the distance' ... what's not to love, Loz?" But - with Pearson-like impatience - we leap ahead. The siren call of the seductive Rufford Arm must be resisted until an inventory is made of the main line's attractions; and they are numerous enough.

Parbold is as pretty as a picture, and it is easy to overlook the historic truth that the Leeds & Liverpool, as initally surveyed, was to take a markedly different direction, northwards, at this point. The widening of the canal between bridges 38 and 37D, is a reminder of the Parliamentary Line, approved in 1770, which would have taken a route to the south of Preston and north of Burnley. In the event the line was never built beyond a drydock. *continued overleaf:*

---

*...along with: Bulbourne, Dukinfield, Hazelhurst, Pelsall, Prees, Saul and Stourton.

for details of facilities in Burscough turn to page 93

*figures relate to main line; allow 2 hours Burscough Junction - Rufford

continued from page 91:

The accompanying countryside is largely flat and much given over to market gardening, with turf-growers providing interesting variety in the vicinity of Bridge 36, one of two electrified swing-bridges. There's a plethora of aqueducts too, variously spanning watercourses or country lanes. Newburgh Aqueduct carries the canal over the River Douglas which, it is tantalising to consider, was navigable in the early years of the canal, so that working vessel crossing working vessel was an everyday occurrence. Indeed, the aqueduct was slightly skewed to facilitate access by river boats into a lock immediately upstream. The Douglas Navigation can trace its origins back as far as 1712, though it was not fully operational for another thirty years. To place this early progenitor of water transport in historical context therefore, it was open when Bonnie Prince Charlie lodged in Wigan during his ill-fated invasion of 1745. Many of the vessels which worked along the navigation were sufficiently sea-worthy to work around the coast to Furness or across the Irish Sea and the boon it brought to the Wigan coalfield was immeasurable.

Visitor moorings are provided by Bridge 34, though the Ring o' Bells pub has closed down. Beyond the towpath a sunken wall gives onto cobbled standing. This was one of many manure wharves in the neighbourhood, where Liverpool's malodorous 'nightsoil' (a euphemism for human excrement) was unloaded to fertilize West Lancashire's market gardens.

If you've persisted with this guide all the way from Dewsbury (Map 4), the maker's name on Bridge 32A will ring a bell. Ainscough's huge flour mill - sympathetically converted into living space - was an important source of traffic: grain coming in from Liverpool; coal from Wigan. The preserved long boats *Ambush* and *Viktoria* belonged to Ainscough's own fleet. Another of their vessels, *Parbold*, was the last horse-drawn boat to ply the canal, carrying coal to the mill until 1960. Incidentally, Hugh and Richard Ainscough, proprietors of the mill, share an elaborate memorial in Our Lady & All Saints churchyard, Parbold.

Burscough was a pivotal point on the canal in its commercial heyday, and to some extent remains so today. Burscough Wharf, by Bridge 32A, was redeveloped in 2011 for a mixture of retail and leisure uses. But in the past there were boatbuilding and repair yards, boatmen's houses, boat horse stables and a veterinary practice devoted to the animals' upkeep. Indeed the canal company even owned a farm which produced provender for the horses.

The Rufford Branch

Far more popular with boaters, now that it leads to the 'Ribble Link', the Rufford Arm was opened in 1781, at a stroke rendering the upper reaches of the Douglas Navigation obsolete. Descending through eight broad (but only 62 feet long) locks, a rural atmosphere is rapidly established as the waterway traverses a landscape reclaimed from marshland early in the 18th century, its peaty soil providing a perfect loam for market gardening. The cut's broad margins are colonised by reed, iris, arrowhead and water lily; the banks burgeon with willowherb, bindweed and bracken. Larks sing overhead whilst coots and moorhens inhabit the reed beds. There are views east towards transmitter-topped Winter Hill - Map 31.

Potato crops are a significant part of the local agricultural economy. Their cultivation hereabouts dates back to the middle of the nineteenth century when the local entrepreneur, James Martland, founder of the Potato Merchants Steamship Company, was proclaimed 'potato king of the world'. His memorial can be found in St John the Baptist's graveyard, Burscough Bridge. Nearing picking time, the potato plants sprout cream or violet coloured flowers. Similarly, the Rufford Arm's locks sprout all sorts of weird, wonderful, and generally user-friendly paddle gear, including a number of cloughs (rhymes with 'cow'). But it seems sad that the paddles, and indeed the swing-bridges, need to be handcuff-locked in what is ostensibly so remote an area.

The Ormskirk-Preston railway accompanies the canal, once an important route of the old Lancashire & Yorkshire Railway, but now a single track byway with a shuttle service of diesel multiple units between Ormskirk and Preston. The very last scheduled steam-hauled passenger train in Britain, the 9.25pm Preston-Liverpool, ran along this line on 3rd August 1968 behind 'Black Five' 45318.

Burscough

Map 34

A cat's cradle of roads and railways - with a canal thrown in for free - Burscough is one of those delightfully infectious communities which canal travel frequently throws up against the odds. Canal warehouses apart, the Gothic parish church and Ainscough's mill are the only buildings of any substance. Yet how you wish you could land up here on a winter Saturday when 'The Linnets' (Burscough FC) are walloping the opposition at Victoria Park.

Eating & Drinking

BLUE MALLARD - Burscough Wharf (Bridge 32A). Tel: 01704 893954. Well-appointed first floor canalside restaurant in refurbished wharf. Open 12-3pm Mon-Sat, 5-9pm Tue-Sat and 12-8pm Sun. L40 5RZ
FAT ITALIAN - Moss Lane (A59). Tel: 01704 895757. Restaurant bar & grill with branches in Ormskirk and Crosby. Open all day from noon. L40 4AY
HOP VINE - Liverpool Road North (Burscough Bridge railway station). Tel: 01704 893799. *Good Beer Guide* listed pub belonging to the Burscough Brewery who occupy premises to the rear. L40 4BY
PURELY PIZZA - Liverpool Road. Tel: 01704 336210. Pizza and pastas: open Wed-Sun from noon. L40 0SA
THE SHIP - Wheat Lane, Lathom (adjoining Lock 2 Rufford Branch). Tel: 01704 893117. *GBG* listed country pub and dining nicknamed the 'Blood Tub'. L40 4BX
SINNER'S CLUB - Burscough Wharf. Tel: 01704 891756. Canalside burger bar open daily ex Mon/Tue. Food from 4.30pm (noon weekends) L40 5RZ

Shopping

There's a Tesco supermarket and Spar convenience store, but a short ride south on a 2A will take you to Ringtail retail park (just off the map) where there's one of Booths' superb supermarkets, a real treat for folk who live beyond their Pennine stomping ground.

Things to Do

MARTIN MERE - 2 miles NW of Burscough. Tel: 01704 895181. Drained wilderness hosting a Wildfowl & Wetlands Trust visitor centre. Signposted footpath from Burscough Bridge railway station. L40 0TA

Connections

BUSES - Stagecoach service 2A runs hourly to/from Preston (via Tarleton and Rufford) and Ormskirk (via Ringtail retail park. Tel: 0871 200 2233.
TRAINS - half hourly (hourly Suns) Northern services from Burscough Bridge to/from Southport (an enjoyable seaside excursion), Parbold, Wigan Wallgate and Manchester. Roughly hour and a half interval Northern service from Burscough Junction station to/from Ormskirk (for Liverpool) and Preston. Tel: 0345 748 4950.
TAXIS - Fremier Taxis - Tel: 01704 896999.

Rufford

Map 34A

Toponymically a 'rough ford' across the River Douglas, the village had strong links with the Hesketh family who owned both the Old and New halls.

Eating & Drinking

ALEXANDER'S BRASSERIE - Liverpool Road (Bridge 8A). Tel: 01704 822040. Hotel restaurant. L40 1SQ
BOAT HOUSE BRASSERIE - St Mary's Marina. Tel: 01704 822458. Cafe/restaurant open daily 9.30am-5pm; last orders 3.30pm. L40 1TD
HESKETH ARMS - Liverpool Road. Tel: 01704 821009. Handsome old coaching inn serving food at lunch and dinner Mor-Sat (ex Wed.) and 1-8pm Sun. L40 1SB
See also Rufford Old Hall tearoom below.
TASTEBUDS - Fettlers Wharf. Tel: 01704 822888. Coffee shop open 9am-5pm daily. L40 1TB

Things to Do

RUFFORD OLD HALL - canalside, though mooring impracticable and official access via entrance on A59. Tel: 01704 821254. House, gardens, shop and tearoom from 11am mid Feb to end Oct daily ex Thur & Fri. Admission charge. Sublime sixteenth century house preserved by National Trust in eighteen acres of gardens and woodland. Shakespeare is said to have performed here for the Heskeths. L40 1SG

Connections

BUSES - Stagecoach service 2A runs hourly to/from Preston (via Tarleton) and Ormskirk via Burscough. Tel: 0871 200 2233.
TRAINS - Mon-Sat Northern shuttle service to/from Preston and Ormskirk (via Burscough Junction). Tel: 0345 748 4950.

Tarleton

Map 34A

A tall spired Victorian church pinpoints this friendly, thriving village which offers a reassuring sense of civilisation in the sequestered wastes and levels which surround the Douglas' descent to the sea.

Eating & Drinking

COCK & BOTTLE - Church Road. Tel: 01772 812258. Thwaites pub open daily from noon. PR4 6UP
GIGI - Mark Square. Tel: 01772 816815. Italian restaurant. Closed Mondays. PR4 6TU
TARLETON TANDOORI - Church Road. Tel: 01772 812200. Indian takeaway from 5pm daily. PR4 6UP
VILLAGE INN - Mark Square. Tel: 01772 815959. Open from noon, food served throughout. PR4 6TU

Shopping

Tarleton has a surprisingly good range of shops including a pharmacy, butcher, delicatessen, bakery and Spar with post office counter. Co-op and Aldi.

Connections

BUSES - Stagecoach service 2 operates hourly to/from Preston and Southport; 2A operates hourly to/from Preston and Ormskirk. Tel: 0871 200 2233.

MARKET gardening gives way to wider, prairie-like expanses of cornfield and grassland. In some ways it reminds you of the Chelmer & Blackwater Navigation in Essex, another backwater which builds to an estuarial climax.

At Rufford, trains between Ormskirk and Preston still go through the time-honoured routine of the driver and signalman exchanging tokens. At Sollom - which is too pretty a hamlet (devoted to race horse training) to live up to the sound of its name - the canal abandons all attempts to continue impersonating a man-made waterway, and joins the original course of the Douglas river beyond the narrows of a former lock.

By Bridge 11 an old warehouse (now used as a fireplace emporium) recalls lost trade. Masked by dense woodland stands Bank Hall, a remarkable 17th century Jacobean mansion, painstakingly being brought back to life after years of dereliction. Architecture lovers will also warm to Becconsall Old Church, charmingly small and brick-built with a bell-cote. Not used regularly since 1926, it is maintained by the admirable Churches Conservation Trust and concerts regularly take place within its enchanting interior.

Prior to creation of the Millennium Ribble Link in 2002, the Rufford Arm petered out at Tarleton; few boaters having the desire, yet alone experience, to lock down into the tidal Douglas. Most of the almost £6 million cost was involved in upgrading Savick Brook to navigable standard, including the construction of nine locks.

Voyages between the Rufford Arm at Tarleton and the Lancaster Canal on the outskirts of Preston are strictly controlled by the Canal & River Trust. Convoys of pre-booked (see page 111) inland waterway craft make the journey in one direction only on days selected to coincide with favourable tidal conditions in the two rivers. Hire boats are not usually permitted onto the Ribble, whilst privately owned craft have to comply with safety conditions - anchors, life jackets (for all crew members) and mobile phones are mandatory. Such stringencies notwithstanding, the passage between the two canals has rapidly become recognised by experienced boaters as one of the most exciting and enjoyable itineraries currently available on the inland waterways system. Purely coincidentally, another can be found on Maps 40a & b!

PUNCHING the incoming tide, you voyage down the River Douglas initially between high flood banks, only beginning to gather speed as the tide slackens and the river widens. The abutments of a swing-bridge which once carried the Preston to Southport railway presage an arcing curve to the east and the premises of Douglas Marine, beyond which you emerge into a marshland environment where sheep and cattle reassuringly graze. Perches count down the miles to the Douglas' confluence with the Ribble. In a setting reminiscent of Trent Falls (where the Trent and Ouse join to form the Humber) you are instructed to steer to port (left) of Asland Lamp before immediately turning to starboard (right) to follow the Ribble upstream.

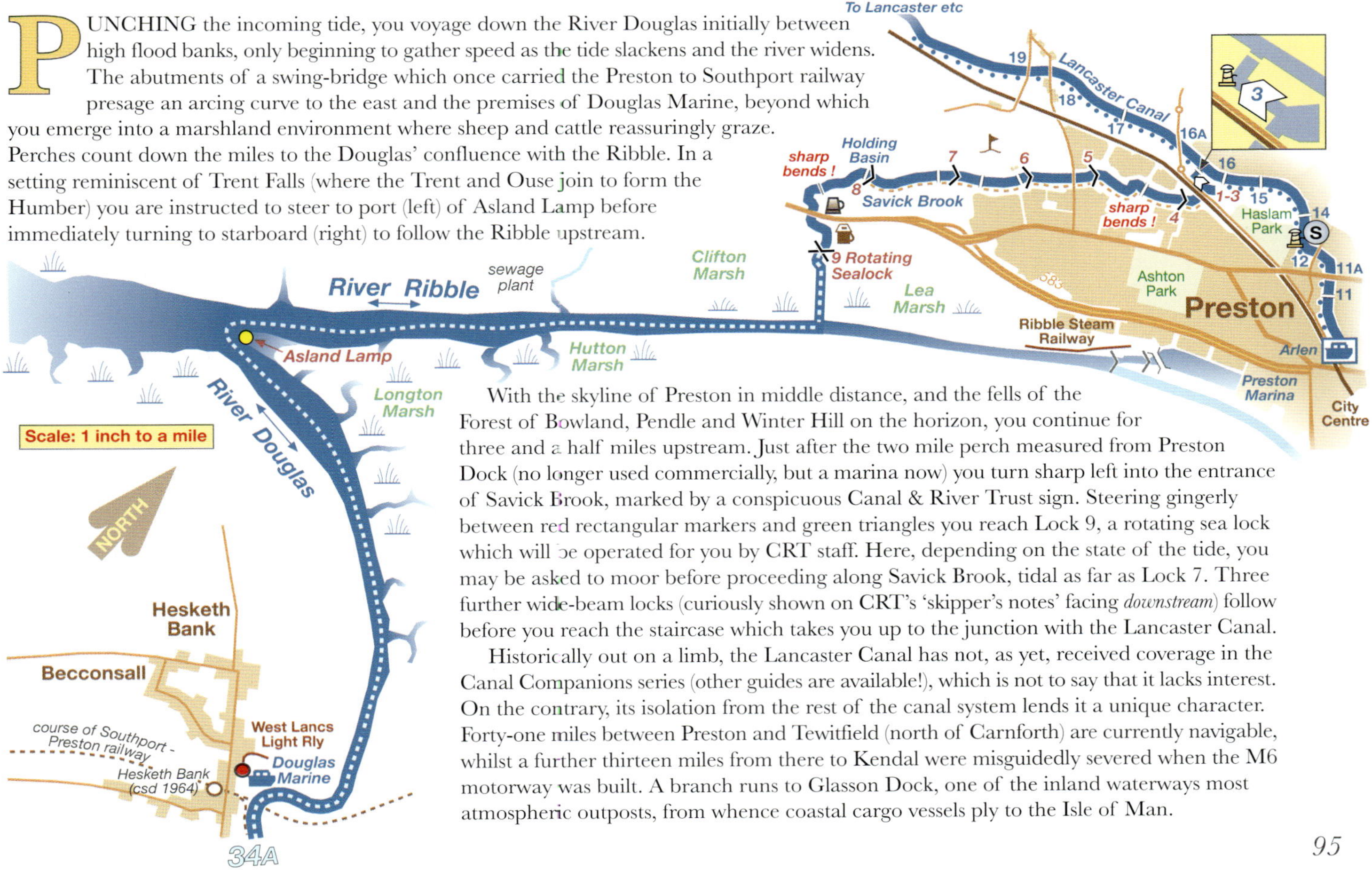

With the skyline of Preston in middle distance, and the fells of the Forest of Bowland, Pendle and Winter Hill on the horizon, you continue for three and a half miles upstream. Just after the two mile perch measured from Preston Dock (no longer used commercially, but a marina now) you turn sharp left into the entrance of Savick Brook, marked by a conspicuous Canal & River Trust sign. Steering gingerly between red rectangular markers and green triangles you reach Lock 9, a rotating sea lock which will be operated for you by CRT staff. Here, depending on the state of the tide, you may be asked to moor before proceeding along Savick Brook, tidal as far as Lock 7. Three further wide-beam locks (curiously shown on CRT's 'skipper's notes' facing *downstream*) follow before you reach the staircase which takes you up to the junction with the Lancaster Canal.

Historically out on a limb, the Lancaster Canal has not, as yet, received coverage in the Canal Companions series (other guides are available!), which is not to say that it lacks interest. On the contrary, its isolation from the rest of the canal system lends it a unique character. Forty-one miles between Preston and Tewitfield (north of Carnforth) are currently navigable, whilst a further thirteen miles from there to Kendal were misguidedly severed when the M6 motorway was built. A branch runs to Glasson Dock, one of the inland waterways most atmospheric outposts, from whence coastal cargo vessels ply to the Isle of Man.

THE beauty of Britain lies in its ability to effect quick costume changes of scenery. Five miles back we were revelling in the woods and meadows of the Douglas Valley. Here it is the arable fields and market gardens of the West Lancs coastal plane which catch the eye. They have been likened to The Fens; could be mistaken for the Somerset Levels; but really deserve to be recognized as an intrinsically definable landscape in their own right.

Hangars on the horizon to the south of abandoned swing-bridge 30 recall the existence of HMS *Ringtail*, a Royal Naval Air Station on the outskirts of Burscough. Opened in 1943, it was deployed as a training centre for torpedo fighter squadrons; Seafires (a naval development of the Spitfire) and Fairey Barracudas, Fireflies and Swordfish being the most commonly resident aircraft. On 6th October 1944 a two-man crew perished in Morecambe Bay after their American built Curtiss Helldiver failed to pull out of a practice dive. Messrs Neville and Turner's resting place is St John the Baptist's adjacent to Burscough Bridge railway station. *Ringtail's* operational days were shortlived. The last squadron left in 1946, though it remained in Royal Navy ownership for a further eleven years. Used primarily as an industrial estate in the intervening years, a new retail development opened in 2014, its centrepiece a Booths supermarket poignantly designed to represent a WWII hangar.

Boating at this end of the Leeds & Liverpool used to be largely restricted to the comings and goings of the Mersey Motor Boat Club. Founded in 1932, they were initially based at Litherland, but after the Second World War they built a club house at Lydiate which is commodious enough to include a dance floor. Resident on such a lengthily lockless canal, they probably needed to work off excess energy in rhumbas and foxtrots. Nowadays they also have moorings at Haskayne and Scarisbrick. In any case, boating has been brisker hereabouts since the Mawdsleys of Fettlers Wharf (Map 34A) developed Scarisbrick Marina in 2009 and the lure of the Liverpool Link began to take effect.

Masked by woodland, Scarisbrick Hall - a school these long years - is the phantasmagorical work of Augustus Pugin and his son Edward. As Peter Fleetwood-Hesketh so tactfully put it in Murray's Architectural Guide to Lancashire of 1955: 'In Charles Scarisbrick, Pugin was fortunate in finding a patron as eccentric as himself'!

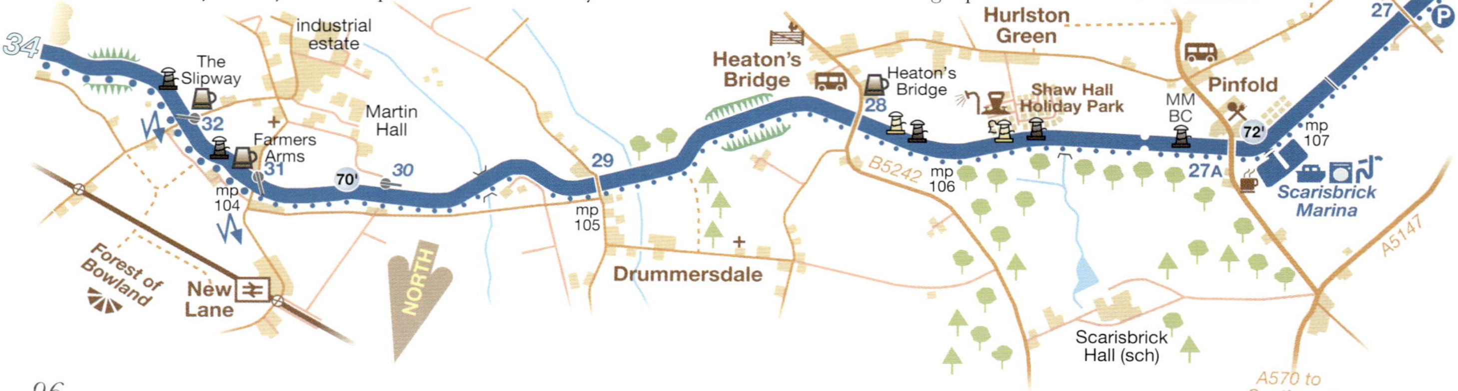

FLAT, yet not for a moment dull, the landscape evokes an air of expectation: after all, the Irish Sea is only half a dozen miles away to the west. So perhaps that sense of intoxication is ozone-fuelled. An absence of rising ground lends itself to extended horizons. To the south-west on clear days the Clwydian Range is clearly defined on the Welsh side of Liverpool Bay; to the north-east the Forest of Bowland rears provocatively up beyond Preston.

By Bridge 25 is Thompson Dagnall's half-buried sculpture of a navvy, sited here to mark the cutting of the 'first sod' on 5th November 1770. Dagnall's other work includes the disgraced 'Ribble Piddler', hastily withdrawn from its site on the Ribble Link in 2008 because it was deemed a danger to the public; not on moral grounds, but because the oak it was hewn from had cracked. A plaque in Halsall Cutting also commemorates where construction of the Leeds & Liverpool began in earnest in 1770. The fact that there is a cutting at all - deep enough to require a retaining wall on the towpath side - suggests that the country-side is not quite as level as it appears to be. Cuttings were not all bad news. If they were rocky it could be used for building bridges; softer strata was employed for the building up of embankments.

The Liverpool Southport & Preston Junction Railway's passenger services were provided by a steam railmotor affectionately known as 'Altcar Bob'. With scant custom in the sparsely populated mosslands threaded by the line, passenger trains were withdrawn before the Second World War. Back in the day, it would have been a rewardingly mischievous exercise to stride into the Booking Hall at, say, Euston, and request of a baffled clerk a First Class Single to Plex Moss Lane Halt. If such things stir your souls as much as they do ours, several overbridges remain intact on by-roads running west from the A5147.

Substantial girders carry the Maghull to Southport *continued overleaf:*

continued from page 97:

road across the canal upon bridges 20A and 21A - as does 17A on Map 37. The big redbrick road-house called the Scarisbrick Arms by Bridge 20A is prominently dated 1899, suggesting that the road was upgraded in that year, and that supplementary bridge numbers had to be added to the original sequence.

Look out for Downholland Hall Farm by disused swingbridge 21. The bulk of the present building predates the canal by a mere forty years, but in previous incarnations it was the venue of the annual Court Leet at which the Lord of the Manor would preside over matters of tenancy. Its great mullioned windows are a particular delight, though Pevsner appears to have been driven by oblivious. He was similarly dismissive of Aughton's high-towered Victorian church which provides a landmark on the eastern horizon: "Large, rock-faced, Dec and not attractive."

Once upon a time, a healthy sibling rivalry existed between Lancashire's industrial powerhouses, Liverpool and Manchester. Then, in the early 1970s, Edward Heath's Conservative government came up with the bright idea of cleaving them both from their parent county, and in doing so created the nebulous territories of Merseyside and Greater Manchester. The two fiefdoms have been inclined to dislike one other intensely ever since. You can ponder on the predisposition of politicians to interfere with everything in their path (and much that isn't) as you cross the spurious Lancashire and Merseyside boundary between bridges 19 and 20.

New Lane — Map 35

A by-road of solitary disposition crosses the canal and the railway on the way to Martin Mere. 'Tin Tabernacle' a few hundred yards south of Bridge 31.

Eating & Drinking

FARMER'S ARMS - New Lane (Bridge 31). Tel: 01704 896021. Open daily from noon. Food lunchtimes and evenings (from 5pm); Sundays 12-6pm. L40 8JA

Connections

TRAINS - approx bi-hourly (not Suns) Northern services to/from Southport etc from wayside halt to north of Bridge 31. Tel: 0345 748 4950.

Heaton's Bridge — Map 35

HEATON'S BRIDGE - Bridge 28. Tel: 01704 840549. Unspoilt canalside pub in building once used as an L&L wharf office. Lunch & evening food. Moorhouse (of Burnley) beer plus guests. Closed Tues. L40 8JG *Also Village Bakehouse Cafe & Bakery.*

Scarisbrick — Map 35

The centre of the village (only three miles from Southport's seaside) lies north of the meeting of the A570 and A5147. Shaw Hall Holiday Park (Tel: 01704 840298 - L40 8HJ) is a caravan & camping centre which offers moorings, a shop, restaurant, and laundry.

Eating & Drinking

SCARISBRICK MARINA - Southport Road (Bridge 27A). Tel: 01704 841924. Cafe open 9am-4pm daily. Good, homely filling food and ice cream. L40 9RH
NELLIE'S - Southport Road (Bridge 27A). Tel: 01704 841333. Indian restaurant (previously called 'The Elephant' - Mandy Miller would approve the derivation!) open from 5.30pm daily. L40 8HQ

Connections

BUSES - Arriva service 300 operates half-hourly (hourly Suns) to/from Liverpool (via Maghull) and Southport (via Scarisbrick). Tel: 0871 200 2233.

Halsall — Map 36

St Cuthbert's 14th century church is approached through a lych-gate flanked by two bulky yews. The Latin inscription on the clock refers to the passage of time and our tendency to do little with it.

Eating & Drinking

THE SARACEN'S HEAD - Summerwood Lane (Bridge 25). Tel: 01704 840204. Whitewashed canalside pub. Lunches and evening meals (from 4.45pm) ex Mon. Food Sundays noon until 6.30pm. L39 8RH

Connections

BUSES - Arriva service 300 operates half-hourly (hourly Suns) to/from Liverpool (via Maghull) and Southport (via Scarisbrick). Tel: 0871 200 2233.

Haskayne — Map 36

Intriguing how age-old farms co-exist so harmoniously with modern housing in all these West Lancs villages. Nice walks to be had westwards across the mosses.

Eating & Drinking

SHIP INN - Rosemary Lane (Bridge 22). Tel: 01704 840077. Canalside pub operated by Lees Brewery of Middleton, Manchester. Daily from noon. L39 7JP
KING'S ARMS - Delph Lane. Tel: 01704 840033. Cosy village local serving up to five real ales. L39 7JJ

Downholland Cross — Map 36

NEW SCARISBRICK ARMS - Black-a-Moor Lane (Bridge 20A). Tel: 0151 526 4004. Substantial Victorian 'road-house' reinvented as a restaurant. Open Mon-Sat for lunch and dinner (from 5pm) and Sunday from noon to 7.30pm. L39 7HX

SUBURBS and countryside arm wrestle for supremacy, and mostly - inevitably - the former gain the upper hand. Post war 'one inch' Ordnance Survey maps tell a different tale, but already depict suburbia's spreading stain. Once upon a time the soundtrack of a journey along this section of the Leeds & Liverpool would have been lark song, now it's ice cream vans; invariably - in our experience - playing *Greensleeves*. Henry VIII will be calculating his royalties as he elbows his way through the queue for a '99' - "make that *three* flakes, knave!"

But someone who wouldn't have readily objected to all these new homes, was a man who saw the opportunity to fill them with toys. Frank Hornby was born to a Liverpool provisions merchant in 1863, and subsequently worked for his father as a cashier. In no way trained as an engineer, he made models for his own sons in his spare time, assembled from metal strips held together by nuts and bolts, a methodology patented in 1901 and eventually marketed as Meccano. Other toys - for boys of all ages - ensued: eponymous clockwork trains, followed by Dinky die-cast model road vehicles. They made him a millionaire and - as millionaires tend to do - he went into politics, representing Liverpool's Everton ward for the Conservatives. He met his wife, Clara, in the chorus of the Liverpool Philharmonic and they set up home in Maghull, initially at *The Hollies* on Station Road where their occupancy is recalled by a Blue Plaque, and later in a larger house called *Quarry Brook*, which is now on the campus of Maricourt High School. Frank Hornby died in 1936, two years before arguably his greatest legacy, Hornby Dublo, was launched. He is buried in St Andrew's churchyard.

Bridge 12A carries the dual-carriagewayed A59 road across the canal at Maghull. In our motorway age such roads have diminished significance, but this route mirrors the Leeds & Liverpool's transpennine journey, extending from Liverpool to York, by way of Preston, Skipton and Harrogate. Perhaps when canal explorers have all deserted us for apps and e-books, Pearson's will find a new niche market, interpreting the old A-roads of England. As part of Seamus O'Donovan's 'Sabotage Campaign', the IRA blew up swing-bridge 15 on 27th July 1939. In the neighbourhood of Melling, the canal enjoys a brief interlude of open countryside - fields of waving grasses and horses grazing, casting a wistful eye towards Aintree's grandstand, prominent on the southern horizon. Many Liverpool-bound boaters prefer to moor overnight by Bridge 10 rather than become embroiled in the city's northern suburbs.

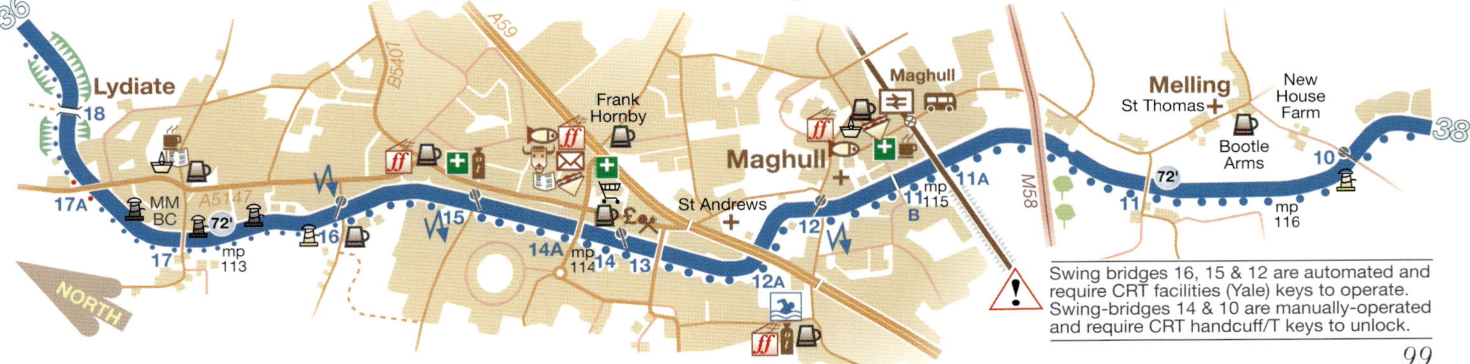

WHEN the Liverpool Link first 'opened' in 2009, boaters were instructed to assemble at Bridge 9 to commence their assisted passage towards Liverpool, British Waterways/CRT employees accompanying them from here onwards to operate the swing-bridges and offer a reassuring measure of security. However, following consultation with interested parties, new procedures were introduced in 2016, effectively providing 'freedom of movement' along the canal north of Stanley Dock Locks (Map 40a). It is important to note, though, that access via the Liverpool Link to the pontoon visitor mooring at Salthouse Dock is only possible between April and October - see page 111.

Canal Turn is one of the most famous horse racing fences in the world. It is the eighth and twenty-fourth jump of the four and a half mile Grand National steeplechase and comes after Becher's Brook and before Valentine's; an illustrious sequence. Having jumped the fence, jockeys face a sharp left-hand turn (for the race is run anti-clockwise) and in the less safety conscious past (for the race has been run here since early Victorian times) it was not unknown for horses to end up in the canal. The fence's most notorious incident occurred in 1928 when a horse called *Easter Hero* refused and in the ensuing commotion twenty other horses and riders fell. By the time the final fence was reached only two horses remained seated and one of those fell, leaving the 100 to 1 outsider *Tipperary Tim*, the winner by process of elimination. Four years earlier, King George V had watched the Grand National from the canal, and there was something of a tradition of boats being decked and decorated as viewing platforms, but in the 1950s a high concrete fence was erected, much to the chagrin of spectators in the habit of watching proceedings gratis. Can you think of any other canalside race course? We could only come up with Wolverhampton and Ripon; Windsor and Worcester being beside rivers. Incidentally, Aintree has also hosted motor racing, the British Grand Prix being held here on a number of occasions in the late Fifties and early Sixties. The Spanish

racing driver Alfonso de Portago is the only person known to have competed at Aintree in both sports, before being killed in horrific circumstances (we'll spare you the gory details) during the Mille Miglia Italian road endurance race of 1957.

Bridge 7C carries the Liverpool to Ormskirk railway over the canal and Merseyrail's third-rail electric trains shuttle by at frequent intervals. It comes as something of a surprise to learn that the Lancashire & Yorkshire Railway introduced electric services along the route before the First World War - many of the north-west's secondary routes were still being electrified a century later. The stone abutments of what must once have been Bridge 7B carried the rival Cheshire Lines Committee's Aintree to Southport route prior to its abandonment in 1952. The suburbs of Netherton and Buckley Hill sidle up to the canal but don't attempt to intimidate it. Three miles to the west lies Crosby and the briny expanse of Liverpool Bay.

Lydiate — Map 37

Echoes of pre-suburban Lydiate straggle the Southport Road (Map 36) in the shape of two Victorian churches. Much older are the ruins of Lydiate Abbey, a victim of the Dissolution of the Monasteries. Just off the map, an isolated thatched inn called - of all things - the Scotch Piper is reputed to be Lancashire's oldest, notwithstanding that it is now in Merseyside.

Shopping
Londis convenience store and newsagent easily reached from Bridge 17.

Connections
BUSES - Arriva service 300 operates half-hourly (hourly Suns) to/from Liverpool (via Maghull) and Southport (via Scarisbrick). Tel: 0871 200 2233.

Maghull — Map 37

Increasingly built-up from the 1930s onwards - so that sometimes you feel like you've unwittingly strayed into Hayes or Harlington - there are still quiet enclaves of discreet bourgeois charm, particularly in the vicinity of the railway station.

Eating & Drinking
FRANK HORNBY - Eastway. Tel: 0151 520 4010. Wetherspoons named after the town's most famous son and featuring examples of Hornby, Meccano and Dinky on display, taking us all back to our childhoods. Open from 10am daily. L31 6BR
GREEK TAVERNA - Liverpool Road North (adjacent Bridge 13. Tel: 0151 531 8640. Open from 5pm daily (noon Sundays). L31 2HB
SCRUMMIES - Station Road. Tel: 0151 345 6209. Thriving daytime cafe opposite the shops near the railway station easily reached from Bridge 12. Open from 7.30am daily (8.30 Sundays). L31 3HF

Shopping
Morrison's supermarket near Bridge 14; additional clusters of shops on Westway and Station Road.

Connections
TRAINS - frequent Merseyrail 'third-rail' electric services to/from Ormskirk and Liverpool. Tel: 0345 748 4950.

Melling — Map 37

Small, unspoilt settlement hanging coyly back from the cacophony of Merseyside. St Thomas dates from 1834 and was formerly known as Holy Rood.

Eating & Drinking
BOOTLE ARMS - Rock Lane (6 minutes walk from Bridge 10). Tel: 0151 526 2886. Marston's 'Two for One' pub. Open from noon daily. L31 1EN

Connections
BUSES - service 133 operates hourly (ex Sun) to/from the railheads at Maghull/Kirkby. Tel: 0871 200 2233.

Waddicar — Map 38

Satellite of Kirkby squeezed between two motorways. Houses occupy the site of a former canalside pottery works. There's a modern pub called the Horse & Jockey (Tel: 0151 739 3601 - L31 1DR) by Bridge 9C and access to a couple of convenience stores, fish & chips (Chinese), a post office and a pharmacy.

Aintree — Map 38

Eating & Drinking
BLUE ANCHOR - School Lane. Tel: 0151 526 1829. Greene King 'Hungry Horse' pub within walking distance of Bridge 9 if you're moored there. Open from 8am-11pm daily, food served throught. L10 8LH

Shopping
Rows of suburban shops within easy reach of bridges 7D and 8.

Things to Do
AINTREE RACE COURSE - Ormskirk Road. Tel: 0151 523 2600. If your canal exploration does happen to coincide with a race meeting you could do worse than attend - if only to see how the other, 'horsey half' live their lives. L9 5AS

Connections
TRAINS - frequent Merseyrail 'third-rail' electric services to/from Ormskirk and Liverpool from Old Roan station near Bridge 7D. Tel: 0345 748 4950.

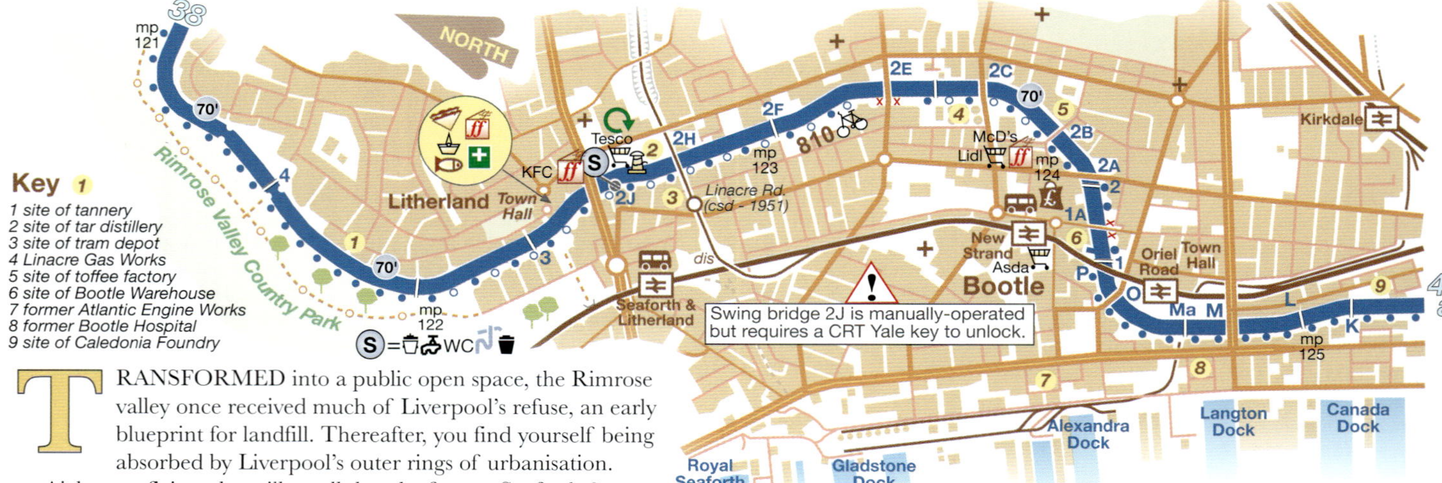

TRANSFORMED into a public open space, the Rimrose valley once received much of Liverpool's refuse, an early blueprint for landfill. Thereafter, you find yourself being absorbed by Liverpool's outer rings of urbanisation.

Aickman aficionados will recall that the former Seaforth & Litherland Constitutional Club came to his rescue, and that of his party, in May 1950 during the IWA's tour of the northern waterways aboard the Wildfowl Trust's converted narrowboat *Beatrice*. A decade on, The Beatles, fresh from a stint in Hamburg, made one of their first appearances, in the neighbouring Town Hall (now a health centre), on the 27th December 1960. John Lennon recalled that most of the audience appeared to be under the illusion that the band *were* German.

Beatrice's progress towards Liverpool had been thwarted by the fact that Litherland's lift bridge didn't operate on Sundays. The lift bridge in question was an ungainly contraption which had replaced an original swing-bridge in 1934 to cope with increased road traffic. But once you pander to road traffic it just keeps growing, and in the mid Seventies a dual-carriageway road bridge was erected. Now, though, the wheel - so to speak - has come full circle and a swing-bridge has been reinstalled, albeit to provide improved pedestrian access to Tesco's supermarket. Boating facilities are provided here, as are offside visitor moorings, protected by security fencing and CCTV, an admirable model. Access to the adjoining Tesco (et cetera) is effected with a CRT Yale key

Seagulls cry over the rooftops of Bootle against a dockland skyline of cranes, masts and funnels. Bootle is a big town with a proud past - witness the dignified, high-spired, clock-towered Town Hall glimpsed from the neighbourhood of Bridge M - but it was from the New Strand shopping centre in 1993 that James Bulger was so terribly abducted. Deprivation and depravity are mutually attracted, but the natives, in our experience, are friendly. Why, even the hoodies say 'excuse me', and at every bridge-hole you half expect to collide with 'the one and only Billy Shears'.

ONE of the Great Waterfronts of the World, Liverpool's old dockland has been accessible to canal boaters since inception of the Liverpool Link in 2009. A 'really good idea' by establishment standards, it is rapidly becoming recognised by the boating fraternity as one of the inland waterway system's holy grails. As dramatic a conclusion to a canal itinerary as crossing the Llangollen Canal's Pontcysyllte Aqueduct, and a good deal more adventurous.

Leeds & Liverpool Canal

Bootle morphs seamlessly into Sandhills. High-walled, the canal keeps its cards closely to its chest. To get in the mood, picture yourself as the cloth-capped, trench-coated, tobacco-chewing steerer of a 'long boat', fresh down from Crooke with coal for the furnaces of Atholl Street Gas Works.

An impressive Leeds & Liverpool Canal company warehouse remains intact by Bridge I. Look out for the 'shipping-hole' through which boats would pass to unload inside the building. On the towpath side stood Hudson's Soap Works. Water was extracted from the canal for filling the tanks and tenders of steam locomotives at the Lancashire & Yorkshire Railway's nearby Bank Hall motive power depot as at Rose Grove (Map 26). Evidence of intense railway rivalry can be gleaned from the former presence nearby of goods depots belonging to the Cheshire Lines, Midland and London & North Western railway companies in addition to the L&Y.

As a counterpart to incoming coal, manure was for many years a staple outgoing cargo, the manure in question being human waste conveyed by boat to the farmlands of West Lancashire for use as a fertilizer. Between bridges H and F the cantilevered loading docks employed to load this revolting cargo onto boats are still there; if not the associated reek. Atholl Street gas works was a recipient of coal by boat until the ice floes of the bitter winter of 1963 brought about an indecent end to the traffic.

The canal reached Liverpool in 1774 - the year before the American War of Independence broke out - and its original terminus was on Old Hall Street. Completion of the route from Wigan was celebrated by (not one but) *two* twenty-one gun salutes. And while the big-wigs partook of a cold collation on the quayside, in excess of two hundred workmen marched by 'with tools on their shoulders and cockades in their hats', following which they were 'plentifully regaled' at a dinner provided on their behalf. The Old Hall Street basins developed rapidly, throwing off additional arms at a dizzying rate, but merchandise for onward carriage by sea still had to be laboriously conveyed by horse & cart between the canal terminus and the docks.

Astonishingly, seventy years were to pass before the canal was linked with the dock system, and it fell to Jesse Hartley - ubiquitous engineer of much of Liverpool's dockland - to design a flight of four locks from a point approximately three-quarters of a mile north of the canal terminus down into Stanley Dock, opened simultaneously with the branch in 1848. With a combined rise/fall of forty-four feet, the chambers were built to a slightly larger gauge than the canal's locks so that Mersey flats (72ft x 14ft 9ins max) could access the canal level. Towards the end of the 19th century, proposals emerged to build a further branch between the canal and the docks at Bootle, but got no further than the drawing board.

Trade at the Liverpool end of the canal complied with the familar pattern of initial success followed by a steady decline as the railway network expanded. But, as in other densely built up parts of the country, factories established along the canal bank continued to use its traffic facilities, particularly for the carriage of fuel from the Wigan coalfield. Henry Tate (native of Chorley and begetter of art galleries) opened his first sugar refinery on Love Lane in 1872, and by the time his business had absorbed Abram Lyle & Sons fifty years later, the works occupied both banks of the canal between Vauxhall Bridge and Chisendale Street. After the plant closed - amidst much recrimination and sadness - in 1981, housing was built on its site, but by that time the canal had been filled in, indeed it had gradually been getting shorter and shorter since the original basin was abandoned in the 1880s. Few vestiges of the original basins remain, save for a (reconstructed) canal office which fronts the Radisson Hotel on Old Hall Street, and a line of impressive warehouses

continued on page 106:

Scale: 4 inches to a mile
NORTH
COMMERCIAL ROAD
VAUXHALL ROAD A5038
BOUNDARY ST
Kingsway Tunnel
CHISENDALE ST
LEEDS STREET
6
LEEDS & LIVERPOOL CANAL
mp 126
E
BLACKSTONE
D
7
C
72'
Vauxhall Bridge
45'
BURLINGTON ST
infilled
Canal Warehouses
4
F
Eldonian Village
8
PALL MALL
H
5
Stanley Dock Locks
1
LOVE LANE
Bank Hall
Sandhills
STREET
2
Sandhills
LANE
A5055
3
Chinese supermarket
B3 (AA)
4
GREAT HOWARD STREET
ventilation tower
grain w'hse
WATERLOO ROAD
2
DERBY ROAD
tea w'hse
East Waterloo Dock
BANKHALL STREET A5056
3
sugar silo
A565
A5054
Titanic Hotel
8
Stanley Dock
tobacco w'hse
REGENT ROAD A5036
Victoria Dock
Jesse Harley Way
39
Subway
B2 (BB)
graving docks
Clarence Dock
Trafalgar Dock
West Waterloo Dock
9
Central Docks Channel
B1 (CC)
REGENT ROAD
Collingwood Dock
b
Wellington Dock
Everton FC (under construction)
Nelson Dock
Salisbury Dock
Liverpool Link
Sandon Half-Tide Dock
Victoria Tower
River Mersey
Huskisson Dock
Canada Dock
Key 1
1 site of Bank Hall mpd (L&Y)
2 Bank Hall Warehouse
3 site of soap works
4 rems of manure wharves
5 site of Huskisson Goods (CLC)
6 former tobacco works
7 site of Athol Street gas works
8 site of Tate & Lyle sugar refinery
9 site of Clarence Dock power station

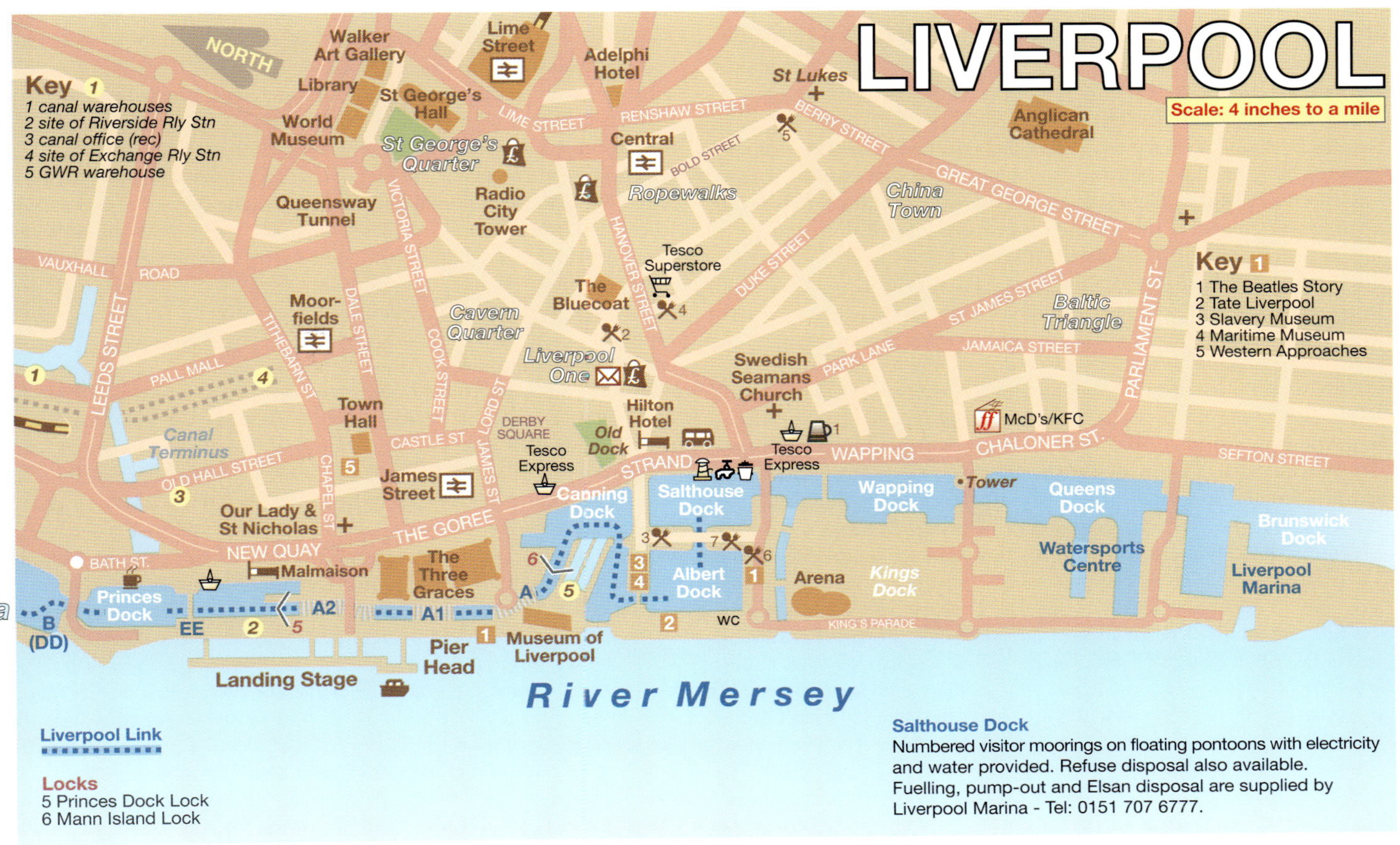
LIVERPOOL
Scale: 4 inches to a mile
NORTH
Walker Art Gallery
Lime Street
Adelphi Hotel
St Lukes
Anglican Cathedral
Library
St George's Hall
World Museum
LIME STREET
RENSHAW STREET
BERRY STREET
Key 1
1 canal warehouses
2 site of Riverside Rly Stn
3 canal office (rec)
4 site of Exchange Rly Stn
5 GWR warehouse
St George's Quarter
Central
Ropewalks
BOLD STREET
China Town
GREAT GEORGE STREET
Queensway Tunnel
Radio City Tower
HANOVER STREET
Tesco Superstore
DUKE STREET
ST JAMES STREET
Baltic Triangle
PARLIAMENT ST.
Moorfields
The Bluecoat
Key 1
1 The Beatles Story
2 Tate Liverpool
3 Slavery Museum
4 Maritime Museum
5 Western Approaches
VICTORIA STREET
DALE STREET
Cavern Quarter
COOK STREET
Liverpool One
Swedish Seamans Church
PARK LANE
JAMAICA STREET
Canal Terminus
Town Hall
LORD ST
JAMES ST
Hilton Hotel
WAPPING
CHALONER ST.
McD's/KFC
SEFTON STREET
OLD HALL STREET
TITHEBARN ST
Tesco Express
Old Dock
Tesco Express
DERBY SQUARE
CASTLE ST
CHAPEL ST
James Street
STRAND
Canning Dock
Salthouse Dock
Wapping Dock
Tower
Queens Dock
Our Lady & St Nicholas
THE GOREE
Watersports Centre
Brunswick Dock
NEW QUAY
BATH ST.
Malmaison
The Three Graces
Albert Dock
Arena
Kings Dock
Liverpool Marina
Princes Dock
A2
A1
A
WC
KING'S PARADE
B (DD)
EE
Pier Head
Museum of Liverpool
Landing Stage
River Mersey
VAUXHALL ROAD
LEEDS STREET
PALL MALL
CHAPEL ST

Liverpool Link

Locks
5 Princes Dock Lock
6 Mann Island Lock

Salthouse Dock
Numbered visitor moorings on floating pontoons with electricity and water provided. Refuse disposal also available. Fuelling, pump-out and Elsan disposal are supplied by Liverpool Marina - Tel: 0151 707 6777.

continued from page 103:

on Pall Mall, retained as a mock frontage to the North Point development. Eldonian Village creates a quiet enclave alongside the remaining stub of the canal and is provided with visitor moorings handily placed for boaters poised to go down the Stanley Flight. Vauxhall Bridge was ceremoniously opened by Cilla Black (and several hundred children) in 1994.

The Liverpool Link

Stanley Locks are open for boats travelling into Liverpool from 1-4pm daily (ex Tue), and for boats travelling out of Liverpool from 8-9.30am daily (ex Tue), and the flight is padlocked outside these times. CRT staff, volunteer or otherwise, may be on hand to assist. Proceeding through a low-head-roomed bridge numbered B3 on its brickwork (though AA on CRT's instruction leaflet) you enter Stanley Dock, framed on its north side by an archetypal Jesse Hartley warehouse imaginatively refurbished as the Titanic Hotel, and a positively gargantuan tobacco warehouse in the process of being redeveloped as apartments on the south side.

A bascule bridge (BB) provides an impressive portal to Collingwood Dock. At Salisbury Dock boaters turn to port (*left*, landlubbers, *left*) gaining, in the process, a close-up of Hartley & Hardwick's hexagonal Victoria Tower, with clock faces on all six sides offering stevedores no alibis in the old days for tardiness. To the north, beyond Nelson Dock, Bramley Moore Dock has been infilled (with sand dredged from the Mersey estuary) and Everton Football Club's new stadium is being erected on the site.

As Liverpool's shipping migrated ever northwards along its waterfront, a number of docks were filled in. Clarence, Trafalgar, and Victoria suffered this fate and an air of despondency filled the void, but work has started on the Liverpool Waters/Central Docks redevelopment scheme with which it is planned to regenerate the whole area, its first manifestation being the Jesse Hartley Way access road which now spans the Central Docks Channel between bridges CC and DD. A new Isle of Man ferry terminal is also planned in the vicinity.

Bona fide boaters apart, the public are currently denied access to this area of the docks. Walkers and cyclists bent on exploration are reduced to following Regent Road; though at least they get to admire the Dock Wall, punctuated at intervals by rotund gateways poignantly displaying the names of the old docks in faded Gothic script. Incidentally, Clarence Dock received in excess of a million migrants of the Irish Famine in the middle of the 19th century, before being filled-in in the late 1920s and a power station erected on the site, its three tall concrete chimneys being referred to, with native Scouser wit, as 'the ugly sisters'.

Liverpool Linkers thread their way tentatively through the buoyed waters of Waterloo Dock - subject, in 1962, of a painting by L. S. Lowry. The West Dock offers tantalising views across the Mersey to Birkenhead, where the domed tower of Wallasey Town Hall is a prominent landmark. The East Dock is dominated by George Lyster's massive grain warehouse of 1868, converted into prestigious apartments. On Waterloo Road a ventilation tower (together with its counterpart on the far side of the Mersey) hints at the subterranean course of Kingsway, newer of two road tunnels beneath the river.

A railway station for boat trains was constructed between Pier Head and Princes Dock. In its hey-day - before the advent of transatlantic air travel - Riverside's platforms were trod by the wealthiest and most glamorous people in the world. Poignantly, the station's approach tracks remain embedded in granite setts beside Bridge DD. Two other railway systems added to the commotion of the city's waterfront. The Mersey Docks & Harbour Board's lines permeated virtually every inch of dockland, expediting transhipment of exported and imported goods to and from the ships at berth. Its elevated counterpart, the Liverpool Overhead Railway, shadowed the riverbank all the way from Seaforth in the north to Dingle, seventeen stations to the south. Affectionately known as the Docker's Umbrella, closure and dismantling of the line's wrought-iron girders in 1957 left scar tissue in the minds of those who had known and loved this unique transit system which has never fully healed. The same harrowing year witnessed the end of the city's trams as well, an extensive ninety mile network which radiated from three loops and a cat's cradle of intersecting tracks at Pier Head. Liverpudlians have a not surprising tendency to paranoia when you consider what has been snatched away from them.

Of incongruous inland waterway modesty in this post-maritime setting, two new locks usher the partially subterranean channel of the Liverpool Link between Princes Dock (whose last commercial use was by the Belfast ferry in 1981) and Canning Dock. Some will consider this the highlight of their passage, passing an equestrian Edward VII, four local musicians who did quite well for themselves, and the 'Three Graces': the Royal Liver Building (topped by its resident birds), the Cunard Building, and the Port of Liverpool Building. Nearby (appropriately facing the Landing Stage) stands the poignant Memorial to the Heroes of the Marine Engine Room. Unveiled in 1916, it commemorates the stark events of April 15th 1912. Though the doomed *Titanic* had sailed from Liverpool's nemesis Southampton, this was its port of registry, and news of the disaster first reached the White Star Line's prestigious (Norman Shaw designed) offices on James Street by telegraph. Fittingly, ocean liners are once again regular visitors to Pier Head, albeit cruise ships, an altogether more sybaritic breed.

Mann Island Lock (where boats leaving Liverpool on their return journey should muster at 8am in order to clear Stanley Locks by 9.30am) provides access to Canning Dock. Two starboard right-angle turns take you into Canning Half-Tide Basin. Half-Tide basins - of which there were a number - were a recommendation of William Jessop's, designed to ease the passage of vessels between the Mersey's capricious tideway and the enclosed docks. Canning Half-Tide Dock is the sole survivor, for much larger, acutely-angled locks became the favoured means of getting increasingly bulkier shipping into and out of the docks.

Before creation of the new canal, Liverpool's docks have always been divided into two discrete systems, the earlier South End Docks and subsequent North End Docks. In 1715 Liverpool could lay claim to the first purpose built dock in the country. At the system's zenith there were seven miles of docks ranged along the east bank of the Mersey, consisting of forty-four distinct basins. It was said that forty per cent of the world's trade was carried in Liverpool ships. The granite used in the construction of Liverpool's docks came from Creetown in Kirkcudbrightshire, south-west Scotland; the same durable material employed in creating the Thames Embankment. Doyen of the dock engineers was the aforementioned Jesse Hartley whose career spanned thirty years from the opening of Canning Dock in 1829 to Canada Dock in 1858. Under Jesse Hartley's successor, George Lyster, expansion continued; necessarily so as sail gave way to steam and tonnages increased exponentially. Lyster's incumbency coincided with the development of transatlantic shipping. Cunard, White Star, Canadian Pacific et al, an illustrious roll call of 'ocean greyhounds'.

A turn to port through a swing-bridge cast at the Haigh Foundry, Wigan takes you into Hartley's most celebrated work, the Albert Dock of 1845, an almost perfectly preserved example of its era whose quadrangle of five-storied, Doric-colonnaded, elliptically-arched warehouses (designed in consultation with Philip Hardwick of Euston Arch fame) have reinvented themselves as a 21st century visitor attraction nonpareil. Not since Saltaire has the transpennine explorer encountered so confident a re-invention of the 19th century. How is it that all the architectural dinosaurs in between have not followed suit? Lack of money, lack of imagination, lack of momentum? Saltaire and Albert Dock illustrate that sloth is no excuse.

Salthouse Dock evokes not so much an anti-climax, more a coda. Previously it boasted warehouses on three sides. Now there's a hint of emptiness, like an unfurnished apartment. Atkinson Grimshaw (Map 14) churned out paintings of it, and as you gently go astern beside the mooring pontoon, you can picture yourself surrounded by the high-masted sailing vessels which his canvases so atmospherically evoke: and how nice to think that he has shadowed you, all the way from Leeds.

Salthouse predates Albert by ninety years. Originally constructed to serve a saltworks, Hartley adapted it to act as Albert's export dock, and at that time there was a direct link into Canning Dock. Thus, incoming vessels could unload at Albert, load at Salthouse and, proceed back out to sea on an anti-clockwise gyratory principal. Outmoded they may be, and suitable only now for play, but these docks were the Felixstowes and Tilburys of their day, and we love them - 'yeah, yeah, yeah' ...

Liverpool Maps 40a & 40b

'If you want a cathedral we've got one to spare', sang the massed choirs of the Anfield Kop in the illustrious Revie/Shankly era when Leeds and Liverpool shared, not only a canal, but a football rivalry every bit as intense as that between Liverpool and Everton. And these two cathedrals define Liverpool, architecturally and denominationally. Work commenced on Sir Giles Gilbert Scott's Anglican Cathedral in 1904 - fourteen years after Liverpool was conferred city status. In certain lights its massive tower looms over Liverpool with the same innate sense of drama displayed by the Christ the Redeemer statue in Rio de Janeiro. It is not without irony that Scott (grandson of George Gilbert of St Pancras station fame, and creator of the classic red telephone kiosk) was a Roman Catholic. Too easterly, in the context of the city centre, to feature on the accompanying map, the Roman Catholic Metropolitan Cathedral is equally remarkable. Edwin Lutyens was commissioned to design the building in 1932. Surmounted by a dome in excess of five hundred feet high - the Anglican Cathedral's tower is a mere three hundred and thirty feet - it would have dwarfed its counterpart. But the Second World War intervened and Lutyens died in 1944, so the design passed to Frederick Gibberd. Post war austerity dictated a less ambitious, though no less unequivocal scheme. Completed in 1967, it rewrote the rules of ecclesiastical architecture. Three additional churches worthy of your curiosity are: Our Lady & St Nicholas which overlooks the Pier Head and has historic associations with the sea and sailors; St Luke's which looms over the top of Bold Street, a bombed out shell reminiscent of the old Coventry Cathedral; and the Swedish Sailors Church (or Gustav Adolf's Kyrka) on Park Lane, an enchanting redbrick design of 1883 which remains a centre for Liverpool's nordic diaspora. Indeed, as you stroll along the streets, it is brought forcibly home to you that, with the possible exception of Hull, nowhere else in England seems as inalienably *foreign* as Liverpool. Both lie alongside wide rivers which perform decent impressions of the sea, and both have carved their inimitable characters from generations of trade with other countries and cultures. So perhaps alien is to be expected. Counter-intuitive it may be, to turn your back on the Mersey, but do so you must if the city centre is to be explored. Gently rising streets will lead you through thoroughfares of surpassing grandeur to St George's Quarter where the eponymous neo-Grecian Hall, World Museum, Library, Walker Art Gallery, Wellington's Column, Centotaph, equestrian statues of Victoria and Albert and Lime Street railway station, combine to effect the kind of concentrated architectural impact that Birmingham, Manchester and Leeds - if, indeed, not London itself - might covet.

Eating & Drinking

BALTIC FLEET (1) - Wapping. Tel: 0151 709 3116. Classic dockland pub revamped for the leisure age and brewing its own beer. Open from noon daily. L1 8DQ
COTE (2) - Paradise Street. Tel: 0151 709 8487. Reliable French chain bistro open throughout the day. L1 3EU
GUSTO - (3) Albert Dock. Tel: 0151 708 6969. Plush Italian chain in dock warehouse setting. Open from noon daily. L3 4AF
HANOVER STREET SOCIAL (4) - Hanover Street Tel: 0151 709 8784. *Good Food Guide* listed brasserie weekdays from noon, weekends from 9am. L1 4AA
ITALIAN CLUB FISH (5) - Bold Street. Tel: 0151 707 2110. Highly rated Italian restaurant on bohemian Bold Street. Open from 10am daily (11am Sun). L1 4JA
LERPWL (6) - Albert Dock. Tel: 0151 909 6241. Fine dining in the shape of plates/tasting menus. 'Clever, high-end food': Sitwell in *The Telegraph Magazine*.
Wed & Thur 12-2.30pm and 6-9pm; Fri & Sat 12-2.30pm and 5.30-9.30pm; Sun 12-6pm. WL3 4AD
MILLER & CARTER (7) - Albert Dock. Tel: 0151 707 7877. Excellent steak house perched on very edge of colonnaded dock. Open from noon daily. L3 4AF
TITANIC HOTEL (8) - Regent Road. Tel: 0151 559 1444. Stanley's Bar & Grill caters for non-residents and you can moor up to a pontoon alongside. L3 0AN

Shopping

Liverpool One is the city's flagship shopping centre and it couldn't be more handily placed for boating visitors. Deeper into the city, however, are the sort of independently owned shops which discerning Pearson users will warm to: the likes of Bold Street, Cavern Walks, and The Bluecoat; the latter housed in the oldest surviving building in the city centre, a beautiful thing in its own right and home to a plethora of craft shops and galleries together with Kernaghan's antiquarian bookshop. Tesco's 'superstore' on Hanover Street is the easiest port of call for victualling the galley. The Post Office is in W.H. Smith at Liverpool One.

Things to Do

BEATLES STORY - Albert Dock. L3 4AD
MARITIME/SLAVERY MUSEUM - Albert Dock. L3 4AQ
MUSEUM OF LIVERPOOL - Pier Head. L3 1DG
ROYAL LIVER BUILDING - Pier Head. L3 1HU
TATE LIVERPOOL - Albert Dock. L3 4BB
WESTERN APPROACHES - Rumford Street. L2 8SZ

Connections

BUSES - Liverpool One's Bus Station is located opposite Salthouse Dock and provides frequent services to most parts of the city. Tel: 0871 200 2233.
TRAINS - Lime Street for inter-city services; Central, James Street and Moorfields for the Merseyrail network. Tel: 0345 748 4950.
FERRIES - Tel: 0151 330 1003. 'Must do' adventure!
TAXIS - ComCab. Tel: 0151 298 2222.

This Guide

Pearson's Canal Companions are a long established, independently produced series of guide books devoted to the inland waterways, and designed to appeal equally to boaters, walkers, cyclists and other, less readily pigeon-holed members of society. Considerable pride is taken to make these guides as up to date, accurate, entertaining and inspirational as possible. A good guide book should fulfil three functions: make you want to go; interpret the lie of the land when you're there; and provide a lasting souvenir of your journeys. It is to be hoped that this guide ticks all three boxes, and quite possibly more besides.

The Maps

There are forty-five numbered maps whose layout is shown by the Route Planner inside the front cover. Maps 1 to 6 cover the Calder & Hebble Navigation between Sowerby Bridge and Wakefield. Maps 6 to 14 the Aire & Calder Navigation between Wakefield and Leeds (with a detour from Castleford to Selby). Maps 14 to 40a/b cover the Leeds & Liverpool Canal and its branches between Leeds and Liverpool.

The maps - measured imperially like the waterways they depict, and not being slavishly north-facing - are easily read in either direction. Users will thus find most itineraries progressing smoothly and logically from left to right or vice versa. Figures quoted at the top of each map refer to distance per map, locks and swing-bridges per map and average cruising time. An alternative indication of timings from centre to centre can be found on the Route Planner. Obviously, cruising times vary with the nature of your boat and the number of crew at your disposal, so quoted times should be taken only as an estimate. Walking and cycling times will depend very much on the state of the towpath and personal stamina.

INFORMATION

The Text

Each map is accompanied by a route commentary placing the waterway in its historic, social and topographical context. As close to each map as is feasible, gazetteer-like entries are given for places passed through, listing, where appropriate, facilities of significance to users of this guide. Every effort is made to ensure these details are as up to date as possible, but - especially where pubs/restaurants are concerned - we suggest you telephone ahead if relying upon an entry to provide you with a meal at any given time.

Walking

The simplest way to explore the inland waterways is on foot along towpaths originally provided so that horses could 'tow' boats. Walking costs little more than the price of shoe leather and you are free to concentrate on the passing scene; something that boaters, with the responsibilities of navigation thrust upon them, are not always at liberty to do. The maps set out to give some idea of the quality of the towpath on any given section of canal. More of an art than a science to be sure, but at least it reflects our personal experiences, and whilst it does vary from area to area, none of it should prove problematical for anyone inured to the vicissitudes of country walking. We recommend the use of public transport to facilitate 'one-way' itineraries but stress the advisability of checking up-to-date details on the telephone numbers quoted, or on the websites of National Rail Enquiries or Traveline for trains and buses respectively. As reliable as we trust this guide will be, the additional use of a contemporary Ordnance Survey Landranger or Explorer sheet is recommended as they are able to present your chosen route in a broader context. Should you be considering walking lengthy sections of towpath over several consecutive days, the dwindling band of Tourist Information Centres can be relied upon to offer accommodation advice.

Cycling

Bicycling along towpaths is an increasingly popular pastime, though one not always equally popular with other waterway users such as boaters, anglers and pedestrians. It is vital to remember that you are sharing the towpath with other people out for their own form of enjoyment, and to treat them with the respect and politeness they deserve.

A bell is a useful form of diplomacy; failing that, a stentorian cough or the ability to whistle tunefully; light operatic extracts go down very well in our experience. Happily, since the inception of the Canal & River Trust in 2012, it is no longer necessary for cyclists to acquire a permit to use the towpath.

Navigational Advice

It's different up north! All the waterways featured in this guide are of the wide beam variety, a whole new ball game compared to the narrow canals of the midlands and south. Which is not to say they're unapproachable, but that you have to go in with your eyes open, expecting the unexpected, and enjoying it.

Calder & Hebble Navigation

A characteristic of the Calder & Hebble is the provision of flood locks or gates at the upstream end of the navigation cuts, constructed to bypass various meanders of the river. Unless the river is in flood, these will be open to passing boats. After heavy rain, however, CRT staff may close them. At such times boats should remain in the safety of the cuts, not passing down on to the river until the colour marker gauge at the tail of the lock is at least yellow, and preferably green.

C&H locks (57ft 6ins x 14ft 2ins) tend to fill very quickly. Ensure you use the ground paddles first, though at some locks you may find only a gate paddle, in which case it should be opened only a little at a time so as to minimise turbulence. Narrowboats of 60 foot length have been known to navigate the C&H successfully by dint of placing themselves diagonally in the lock chambers.

'Handspikes' are a curious feature of many C&H locks, not some diabolical instrument of torture left over from the Spanish Inquisition, but rather, a piece of timber (not unlike a baseball bat) which you insert into a cog and make a series of manual wrench-like movements to raise the paddle. The handspikes themselves can be obtained from boatyards in the vicinity or home-made using hardwood cut to a suitable shape and dimension.

Aire & Calder Navigation

Built for barges, the A&C is barely used commercially now. Frankly, in its huge automated locks, pleasure craft look preposterous, like toys in the bath. Operation is via a control panel accessed by CRT facilities/Yale key. Follow the instructions sequentially and all should be well. Push button controls are certainly a relief from the heavy manual locks encountered elsewhere in this guide. One quibble! It isn't easy to determine when the push buttons are 'lit'. In daylight they suffer from the same problems which are often encountered at cash machines. Should there be a malfunction, you will need to telephone CRT on 0303 040 4040. Bear in mind, too, that this is a 'navigation' with sections of river. Never proceed onto river sections when the gauge shows only red.

River Aire & Selby Canal

Maximum dimensions between Knottingley and Selby are 78ft x 16ft 6ins. There are no particular problems associated with this route, but the river section between Bank Dole Lock and Haddlesey Flood Lock can suffer from excessive current speeds when the water is high. Boaters should not enter the river when the gauges are red. Send your strongest crew members to operate the manual locks, the gates are somewhat cumbersome.

Selby Lock

24 hours notice must be given to Selby's lock keeper (Tel: 01757 703182) by boaters intent on using Selby Lock to access the tidal Ouse. The lock may be negotiated between two and a half hours before high water and five hours thereafter. It is preferable that crews should consist of a minimum of two and that they should have access to a VHF marine band radio.

Leeds & Liverpool Canal

The chief characteristics of this famous canal are its wide beam locks (62ft x 14ft east of Wigan - 72ft x 14ft Leigh to Liverpool) and its swingbridges. Both have a reputation for being hard graft, though that proviso apart, we encountered no particular problems other than the frustration of having to unlock and relock the vandal-proof 'handcuff' locks attached to a good number of the lock paddles and the majority of the swingbridges.

The Leeds & Liverpool's broad locks are able to assimilate two narrowboats side by side, and as the majority of craft using this canal these days seem to fall into this category, it makes sense to share locks where feasible, both for the sake of saving water and also from the aspect of saving energy.

There are quite a few rise or 'staircase' locks at the eastern end of the canal. Many are overseen by full time or volunteer keepers, but if not you will need to remember to ensure that going uphill the chamber above is full so that it can be emptied into the one you are in, whilst downhill the chamber below you should be empty so as to allow the water from the one you are in to empty into it. All very Archimedean!

Turbulence can be a problem in wide locks when travelling uphill. Some boaters prefer to use ropes (looped around the mid-lock bollard) to steady their craft. Alternatively, draw the paddle on the same side as the boat first, which has the effect of bouncing off the opposite lock wall and holding the boat more or less against the side it's on. Whatever your preferred method, use the ground paddles first until the lock is at least half full. Mention of ground paddles brings to mind the weird and wonderful variety of gear still extant on the Leeds & Liverpool. Don't be nervous of them, survivals like cloughs (pronounced to rhyme with the word 'how'), box paddles and gate 'scissors' are to be cherished in this age of uniformity.

Lastly a few notes regarding specific lock flights: The 3-rise 'staircase' flights at Forge and Newlay (Map 15) have restricted operating times from 8am to 4pm to combat hooliganism and staff are usually on hand to oversee boat passages. The same operating times apply to Bingley's 3 and 5-rise flights (Map 17). The lock flights at Gargrave (Map 21), Bank Newton (Map 22), Greenberfield (Map 23) Barrowford (Map 24) and Johnson's Hillock Locks (Map 30) are heavy going by conventional inland waterway standards, yet present no undue challenge. Blackburn Locks (Map 28) might 'bark', but if you're methodical and unhurried they won't 'bite'. Wigan Locks (Map 32) are undeniably gruelling (and top gates have a tendency to leak) but no longer feature assisted passages. Three to four hours is considered a good time for an able-bodied crew. Double that if you're single-handed. West of Wigan the locks are generally user-friendly.

Swingbridges on the Leeds & Liverpool fall into three distinct categories. Most are manually operated by brute strength, a few require the use of a windlass, whilst others are electrified and controlled from a push-button console accessed with the CRT facilities/Yale key. The moral is to ensure that each stage of the sequence is completed and that the bridge has fully returned to its closed position before attempting to extract the key.

Ribble Link

An increasingly popular means for boat owners to reach the otherwise isolated Lancaster Canal. Passages *must* be booked online in advance. Maximum craft dimensions are 62ft x10ft 7ins. Visit the Canal & River Trust online (where a *Skippers Guide* can be downloaded) or telephone them on 0303 040 4040 for up to date information.

Liverpool Link

An exciting means of reaching the centre of Liverpool by boat. Passages *must* be booked online in advance. Maximum craft dimensions are 72ft x 14ft 3ins. Visit the Canal & River Trust online (where a *Skippers Guide* can be downloaded) or telephone them on 0303 040 4040 for up to date information.

Mooring on the canals featured in this guide is per usual practice - i.e. on the towpath side, away from sharp bends, bridge-holes and narrows. Theoretically, you can moor anywhere - as long as the foregoing stipulations are taken into account - but in recent years budget cuts have resulted in shallow edges and increased vegetation, rendering it difficult to moor at whim. The Canal & River Trust signpost designated visitor mooring sites, often with time limitations to dissuade lingering. A yellow-tinted bollard symbol on the maps represents such sites; though occasionally we also include this symbol if we feel a location especially lends itself to short term mooring but isn't officially designated. On rivers it is advisable to turn your bow to face the current when mooring.

Turning points on the canals are known as 'winding holes'; pronounced as the thing which blows because in the old days the wind was expected to do much of the work rather than the boatman. Winding holes capable of taking a full length boat of around seventy foot length are marked where appropriate on the maps. Winding holes capable of turning shorter craft are marked with the approximate length. It is of course possible to turn boats at junctions and at most boatyards, though in the case of the latter it is considered polite to seek permission to do so.

Boating facilities are provided at fairly regular intervals along the inland waterways - though not as frequently 'up north' as in the midlands and south - and range from a simple water tap or refuse disposal skip, to the provision of sewage disposal, showers and laundry. Such vital features are also obtainable at boatyards and marinas along with repairs and servicing.

Closures (or 'stoppages' in canal parlance) traditionally occur on the inland waterways between November and April, during which time most of the heavy maintenance work is undertaken. Occasionally, however, an emergency stoppage, or perhaps water restriction, may be imposed at short notice, closing part of the route you intend to use.

Canal & River Trust

The Canal & River Trust control the bulk of the inland waterways network. Their Head Office is located - rather unromantically - at: First Floor North, Station House, 500 Elder Gate, Milton Keynes MK9 1BB Tel: 0303 040 4040 *www.canalrivertrust.org.uk*

Societies

The Inland Waterways Association was founded in 1946 to campaign for the retention of the canal system. Many routes now open to pleasure boaters may not have been so but for this organisation. Membership details, together with details of the IWA's regional branches, may be obtained from: Inland Waterways Association, Island House, Moor Road, Chesham HP5 1WA. Tel: 01494 783453. Calder Navigation Society *www.calderns.org.uk* Leeds & Liverpool Canal Society *www.leedsandliverpoolcanalsociety.co.uk*

Amendments

Updates to current editions can be found on our website: *www.jmpearson.co.uk*. Feel free to email us if you spot anything worth notifying others about.

Acknowledgements

This guide was compiled with a little help from the following friends: Peter Appleyard, Terry Bates, Barbara & Richard Bradburn, Peter de Courcy, Lynne Delaney, Justine Doughty, Revd Bill Henderson, David Hymers, Alice Kay, Shirley Levon, David Lewis, Tamar, Tom & Arlo and Oli Lumsden, Stephanie Lyons, Alan & Daniel Mawdsley, Dave Murray, Terry & Christine Rigden, Ken Rumney, Chris Sherburn, Nigel & Susan Stevens, Jenny Tyte, Paul Waddington, and Dave Whyles; a lot of help from Meg Gregory and Annie Heron; and a 'lorra lorra' help from Karen Tanguy. Much appreciation to the Short Run Press of Exeter for their expertise in turning the foregoing material from file to print. Eden Pearson organised his parents' whistle-stop tour of the north for this update, providing classy accommodation in Malmaison hotels en route.

BOATING DIRECTORY

Hire Bases

BEAR BOATING - Apperley Bridge Marina, Leeds & Liverpool Canal, Map 16. Tel: 0796 990 1383. BD10 0UR www.bearboating.co.uk

CANAL BOAT CRUISES - Riley Green, Leeds & Liverpool Canal, Map 29. Tel: 01254 202967. PR5 0SP www.canalboatcruises.co.uk

ELLERBECK NARROWBOATS - Heath Charnock, Leeds & Liverpool Canal, Map 31. Tel: 01257 480825. PR7 4DE (Day Hire Only)

HAPTON VALLEY BOATS - Reedley Marina, Leeds & Liverpool Canal, Map 25. Tel: 01282 771371. BB12 0DX www.waterways-great-britain.co.uk

LOWER PARK MARINA - Barnoldswick, Leeds & Liverpool Canal, Map 23. Tel: 01282 815883. BB18 5TB (Day Hire Only).

PENNINE CRUISERS - Skipton, Leeds & Liverpool Canal, Map 20. Tel: 01756 795478.

BD23 1LH www.penninecruisers.com

SHIRE CRUISERS - Sowerby Bridge, Calder & Hebble Navigation, Map 1 (plus outstation on Leeds & Liverpool Canal at Barnoldswick, Map 23) Tel: 01422 832712. HX6 2AG www.shirecruisers.co.uk

SILSDEN BOATS - Silsden, Leeds & Liverpool Canal, Map 19. Tel: 01535 653675. BD20 0DE www.silsdenboats.co.uk

SKIPTON BOAT TRIPS - Skipton, Leeds & Liverpool Canal, Map 20. Tel: 01756 790829. BD23 1LH (Day Hire Only) www.canaltrips.co.uk

SNAYGILL BOATS - Skipton, Leeds & Liverpool Canal, Map 20. Tel: 01756 795150. BD20 9HA www.snaygillboats.co.uk

WHITE ROSE - Leeds, Leeds & Liverpool Canal, Map 14. Tel: 0771 086 3124. LS12 2DS www.whiterosecanalholidays.co.uk

Boat Yards

APPERLEY BRIDGE MARINA - (Calder Valley Marine) Apperley Bridge, Leeds & Liverpool Canal, Map 16. Tel: 01274 616961. BD10 0UR

BOTANY BAY BOATYARD - Chorley, Leeds & Liverpool Canal, Map 30. Tel: 0796 738 0464. PR6 9AE

CALDER VALLEY MARINE - Dewsbury, Calder & Hebble, Map 4. Tel: 01924 467976. WF12 9BD

CASTLEFORD BOAT YARD - Castleford, Aire & Calder, Map 8. Tel: 01977 513111. WF10 2LG

DB MARINE - Selby, Selby Canal. Map 11. Tel: 0777 098 7716. YO8 8NA

FALLWOOD MARINA - Bramley, Leeds & Liverpool Canal Map 15. Tel: 0113 258 1074. LS13 1ER

FETTLERS WHARF - Rufford Arm, Leeds & Liverpool, Map 34A. Tel: 01704 821197. L40 1TB

HAINSWORTH'S BOATYARD - Bingley, Leeds & Liverpool, Map 18. Tel: 01274 565925. BD16 4DR

LIVERPOOL MARINA - Liverpool, Map 40B. Tel: 0151 707 6777. L3 4BP

LOWER PARK MARINA - Barnoldswick, Leeds & Liverpool, Map 23. Tel: 01282 815883. BB18 5TB

MIRFIELD MARINA - Mirfield, Calder & Hebble, Map 3. Tel: 0781 128 3618. WF14 8NL

PB MECHANICAL SERVICES - Heath Charnock, L&L, Map 31. Tel: 01257 474422. PR7 4DE

PENNINE CRUISERS - Skipton, Leeds & Liverpool Canal, Map 20. Tel: 01756 795478. BD23 1LH

REEDLEY MARINA - Burnley, Leeds & Liverpool Canal, Map 25. Tel: 01282 450531. BB12 0DX

RODLEY BOAT CENTRE - Rodley, Leeds & Liverpool Canal, Map 16. Tel: 0113 257 6132. LS13 1LP

SCARISBRICK MARINA - Scarisbrick, Leeds & Liverpool, Map 35. Tel: 01704 841924. L40 9RH

SELBY BOAT CENTRE - Selby, Selby Canal, Map 11. Tel: 01757 212211. YO8 8NB

SHEPLEY BRIDGE MARINA - Mirfield, Calder & Hebble, Map 3. Tel: 01924 491872. WF14 9HR

SHIRE CRUISERS - Sowerby Bridge, Calder & Hebble Map 1. Tel: 01422 832712. HX6 2AG

SILSDEN BOATS - Silsden, Leeds & Liverpool Canal, Map 19. Tel: 01535 653675. BD20 0DE

SNAYGILL BOATS - Skipton, Leeds & Liverpool Canal, Map 20. Tel: 01756 795150. BD20 9HA

STANLEY FERRY MARINA - Stanley Ferry, Aire & Calder, Map 6. Tel: 01924 201800. WF3 4LJ

ST MARY'S MARINA - Rufford Arm, Leeds & Liverpool, Map 34A. Tel: 01704 823697. L40 1TD

TARLETON BOATYARD - Rufford Arm, Leeds & Liverpool, Map 34A. Tel: 01772 812250. PR4 6HD

WHEELTON BOATYARD - Wheelton, Leeds & Liverpool, Map 30. Tel: 01254 831475. PR6 8EX

WHITE BEAR MARINA - Adlington, Leeds & Liverpool Canal, Map 31. Tel: 01257 481054. PR7 4HZ